D1295628

Adrian E. Scheidegger

Systematic Geomorphology

Springer-Verlag Wien New York

Prof. Dr. Adrian E. Scheidegger
Professor of Geophysics
Technical University of Vienna, Austria

With 251 Figures

Library of Congress Cataloging-in-Publication Data. Scheidegger, Adrian E., 1925— . Systematic geomorphology. Bibliography: p. Includes index. 1. Geomorphology. I. Title. GB401.5.S34. 1987. 551.4. 87-16417

ISBN 3-211-82001-9 Springer-Verlag Wien-New York
ISBN 0-387-82001-9 Springer-Verlag New York-Wien

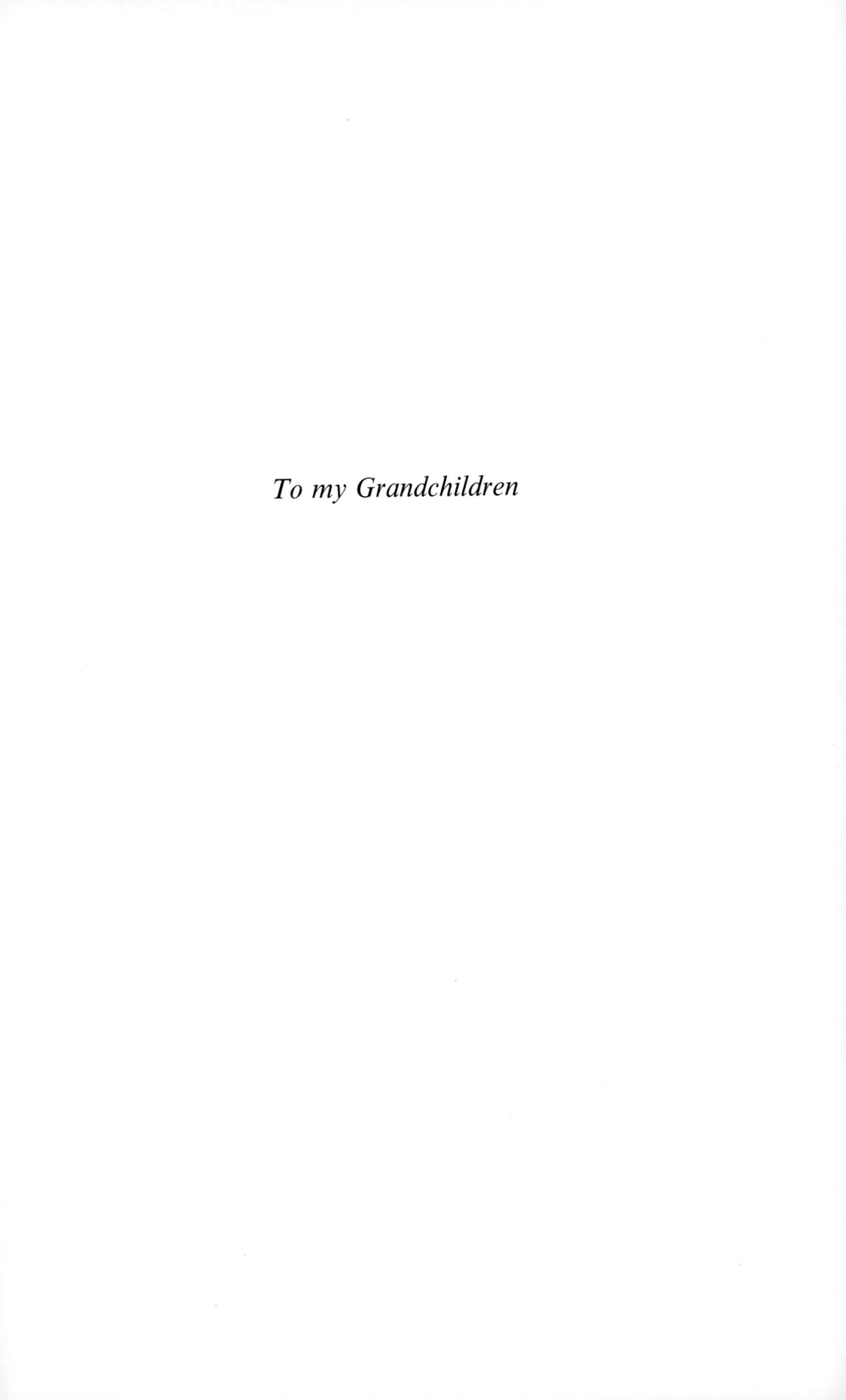

To my Grandchildren

Preface

To most people, travel is an exciting experience. When one journeys around the world, one is struck by the great variety and beauty of the landscapes that one encounters.

The scientific mind, naturally, is not satisfied with admiring the various landscapes, but would like to understand how they were formed. The exact theory of landscape formation is a very complicated affair, but much can be learnt from accurate observation.

The need for the present little book became apparent to the writer during his studies of the mechanics of landscape formation. It turned out that there was, in fact, no systematic compilation of those surface features of the Earth available, that have to be explained by theory. In effect, even the taxonomic principles that have to be applied in a classification of landscapes have nowhere been clearly stated. Thus, this book is intended to present a pictorial taxonomy of geomorphic features based on the basic principles of landscape genesis, as they have recently been worked out.

The pictures have all been taken by the writer himself during many geoscientific studies and travels throughout the world. Some of these pictures had already been used in earlier publications of the writer's. Such previously used pictures have been duly referenced; the writer wishes to acknowledge his indebtedness to the Zeitschrift für Geomorphologie, to Rock Mechanics, to the Gesellschaft der Geologie und Bergbaustudenten in Österreich, to Physik in unserer Zeit, and to the Institut für internationale Architektur-Dokumentation (Arcus) for the permission to republish them.

Vienna, June 1987 **A. E. Scheidegger**

Contents

1 Introduction 1

Taxonomy of geomorphic features in tabular form . . . 7

2 Basic landscape types 11

3 Structural background 27

4 Petrological background 57

5 Slope development 95

6 River action 131

7 Large bodies of water 179

8 Niveal features 213

9 Desert features and related phenomena 239

10 Volcanic landscapes 253

References 275

Index 277

1 Introduction

Systematic geomorphology is the science of the classification of landscapes. Inasmuch as the surface features of the earth are of a bewildering complexity, the task of classifying them is not an easy one.

Traditionally, this task has been tackled on the basis of some evolutionary hypothesis. Thus, Davis (1924) noted that erosion and degradation by water, ice and wind represent some of the most effective agents in landscape evolution. However, since "erosion" can only have a destructive effect in a landscape, he assumed that all development would involve the degradation of forms that had been built up previously by endogenic (tectonic) processes. Thus, Davis supposed that every landscape would be passing through progressive stages of youth, maturity and old age, after some cataclysmic geological event had initiated this process. For each one of these stages, specific characteristics were postulated; thus in "youth", valleys were supposed to be narrow and steep, in "maturity", broad and gentle, and in "old age" all that would remain was supposed to be a plain, corresponding to the base level of erosion. A new "cycle" of landscape evolution would be initiated after a new tectonic cataclysm would create a new topographic relief. Characteristic cycles had been postulated according to whether the climate was humid, glacial or arid.

Unfortunately, the cycle theory of Davis contains a fundamental misconception. None of the evolutionary landscape cycles could ever be followed through on a single object. "Young"-looking forms are seen in one place, "mature" ones in another, "old-age" ones yet elsewhere. The very existence of "cycles" as postulated by Davis is therefore to be questioned. If a classification of landscapes is to be attempted, this can evidently not be done on the basis of an evolutionary theory that is patently false.

Therefore, if a taxonomy of landscapes is to be established on the basis of their genesis, the latter must first be described correctly.

Davis (1924) was certainly correct in his view that erosion

(exogenic) and tectonic (endogenic) processes are involved in landscape evolution. It is also clear that the primary initiation of landscape dynamics has to come from endogenic processes, inasmuch as exogenic degradation (erosion) can only occur if a relief is already in existence.

Nevertheless, endogenic uplift and exogenic degradation do not occur in sequence as supposed by Davis, but concurrently. In this manner, a landscape represents a "system" affected by two antagonistic processes: tectonic build-up and exogenic degradation. If a landscape shows any sort of permanent character, these two antagonistic processes are in dynamic equilibrium ("*principle of antagonism*"; Scheidegger, 1979).

In the light of the above remarks the concepts of "youth", "maturity" and "old age" introduced by Davis for the classification of landscapes, attain now a new significance: If the activity level of the two antagonistic processes mentioned above is high, the landscape has the character of "youth", if the activity is medium, the landscape has the character of "maturity", and if the activity is low, the landscape shows the features of "old age" in the terminology of Davis.

In the above qualitative description, quantitative values have to be assigned to the concept of activity level as well as to the concept of "landscape character".

The activity level is best described in terms of tectonic uplift rates, which must equal the exogenic denudation rates in the case of dynamic equilibrium. The former can be measured by repeated precise levelling operations, the latter by determining the total load carried by a river per unit time at a certain point and dividing it by the total area drained above the point. Experience has shown that there is indeed a rough balance between these two rates (cf. Scheidegger, 1987); in humid, high mountain areas ("high activity"), they are of the order of (nearly) cm/year; in low mountains, they are of the order of mm/year; and in plains regions, they are much below one mm/year.

The second concept, that of landscape "character", was quantified by Strahler (1957) through the introduction of the *hypsometric curve*. The latter is obtained by calculating the fraction of the area under consideration that lies below a certain height-level. In geomorphology, it is customary to consider *relative* hypsometric curves: the heights and areas are divided by the total height (difference between the highest and lowest point in the area) and by the total area under consideration, respectively. Then, Strahler (1957) had shown that the hypsometric curves are convex for "youthful" landscapes, more or less straight for "mature" landscapes and concave for "old age" landscapes. This qualitative statement has been further quantified (Scheidegger, 1987) by the introduction of a "hypsometric index": This is the quotient of the area under the hypsometric curve and the area of the isosceles triangle obtained by drawing a straight line from the point (0, 1) to the point (1, 0) in the hypsometric graph.

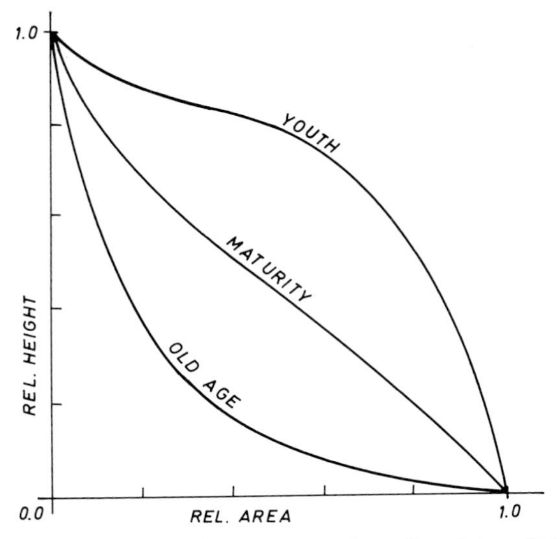

Fig. 1. Hypsometric curves after Strahler (1957)

Now, one has two quantitative measures for describing the landscape dynamics: The uplift/denudation rate, and the hypsometric index.

In view of the above remarks the terms "youth", "maturity" and "old age" should be abandoned in the characterization of landscapes and should be replaced by "high-activity", "medium activity" and "low-activity"-type landscapes. Clearly, one has a hypsometric index of between 2 and 1 for high-activity ("youth"), of around 1 for medium-activity ("mature"), and below 1 for low-activity landscapes (see Fig. 1).

Furthermore, it has been remarked (Scheidegger, 1979) that the two antagonistic processes active in landscape dynamics have a different character: Since tectonic (endogenic) processes act in consequence of plate tectonics, their statistical character is *systematic* over large (plate-size) regions. In contrast, since exogenic processes have their origin essentially in atmospheric processes that are best described by the statistical theory of turbulence, they present the aspect of being stochastic processes on a small scale. The result is that exogenic processes produce features that are *random* (such as meander trains, random drainage nets etc.) on quite small scales. Consequently, the statistical character of geomorphic features can be used in order to obtain an indication of the actual origin of such features.

The configuration of forms in a landscape can generally be characterized by a set of parameters (such as hypsometric index, drainage density, mean slope angle etc.). Inasmuch as a particular configuration in a landscape is the result of a dynamic process, it stands to reason that, if one or more of the parameters are changed, the others must "respond" for there to be an equilibrium. In this connection, a large literature has sprung up discussing and describing the process-responses that take place in the wake of natural and anthropogenic changes in a landscape (cf. Terjung, 1982; Scheidegger, 1988).

In general, the processes respond to changes in the parameters in a gradual way. However, at times, the reponse may lead to an instability.

One reason for an instability to occur may lie in the existence of a positive feed-back mechanism: if a parameter is slightly changed, the adjusted process may further change this same parameter in the same direction. The initiation of this process may lie in natural statistical fluctuations in the system (Taylor instability).

Another reason for the occurrence of an instability of the system may lie in a multivaluedness of the dynamic equilibrium parameters: Thus, the state of the system may suddenly change from one possible equilibrium configuration to another. Furthermore, the various branches of the curves that represent equilibrium states in parameter space can possess singularities; if one of these is approached, an instability threshold is reached (a "catastrophe" occurs; cf. the catastrophe theory of Thom, 1972).

The effect of a Taylor-type instability expresses itself in the fact that geomorphic deviations from uniformity tend to grow. Thus, meanders become bigger (until they are cut off), erosion cirques grow etc. This type of evolution has been described as the consequence of the operation of an "*instability principle*" (Scheidegger, 1983) in geomorphology.

The existence of multivalued dynamic equilibrium parameters expresses itself, for instance, in the presence of reaches with streaming and with shooting flow in a river, in the presence of flat-steep-flat sections on a slope etc. This has been called the "*catena principle*" in geomorphology (Scheidegger, 1986).

We have seen that endogenic processes are the primary agents in geomorphology: Without tectonic activity, no degradation can take place. Thus, there is a "structural background" to all landscapes, which expresses itself in the systematic nature of many features (Scheidegger and Ai, 1986).

In addition to the primary systematism due to endogenic control, there are, however, additional systematic features that are the result of a certain directedness of the exogenic processes. This directedness has been considered as the result of the operation of a "*selection principle*" (Gerber, 1969) in geomorphology which states that the processes of degradation and erosion occur preferentially in such a fashion that statically stable forms are "selected" by them. The stability, of course, is with regard to the stresses induced by the weight of the developing forms themselves. Because the gravitational field is also homogenous over large areas, like the tectonic stress field, the resulting forms equally show a certain systematism. Specifically, the evolution of triangular peaks, towers and similar features can be ascribed to the operation of the selection principle.

From the discussion given above it should become clear that it is not possible to give a "linear" classification of geomorphological features. Since there are several criteria possible as a basis for classification, the result must be a (multidimensional) matrix.

In the light of the principle of antagonism, the most fundamental classification is evidently obtained according to the activity level. Depending on the prevailing climatic conditions, specific fundamental landscape types are the result.

Furthermore, the type of tectonism and of the material present leads to a consideration of the structural and petrological background of a landscape.

Next, specific systems can be considered, of which slopes are the most fundamental ones. In slopes, the operation of the various principles of landscape evolution afford a further classification.

Specific types of landscapes can then be considered individually, depending on their mode of genesis. Thus, chapters will be devoted to fluvial effects, to the ocean-land system, to glacial geomorphology, to desert features and, finally, to volcanic landscapes.

To sum up, the taxonomy of geomorphic features can be presented in form of a systematic table, to which the pictures in this book are keyed. This table is given as Table 1 on the following pages.

Table 1. Taxonomy of geomorphic features in tabular form (figures are keyed to this table)

1 Global aspects 1

2 Basic landscape types 11

2.1 High-activity landscape 12
2.11 Humid climate 12
2.12 Glacial climate 14
2.13 Arid climate 16

2.2 Medium-activity landscape 16
2.21 Humid climate 16
2.22 Glacial climate 19
2.23 Arid climate 20

2.3 Low-activity landscape 20
2.31 Humid climate 20
2.32 Glacial climate 22
2.33 Arid climate 24

3 Structural background 27

3.1 Continuous deformation 28
3.11 Tectonic folds 28
3.12 Folds from differential movement 32
3.13 Instability phenomena 34
3.14 Deformation of petrofabrics 36

3.2 Discontinuous deformation 38
3.21 Joints 38
3.22 Faults 42
3.23 Thrust sheets 45

3.3 Surface effects 48
3.31 Wall effects 48
3.32 Exfoliation 52

3.4 Self-gravitational effects 54
3.41 Mountains 54
3.42 Plateau edge 56

4 Petrological background 57

4.1 Volcanics 58
4.11 Volcanic flows 58
4.12 Volcanic ejecta 62
4.13 Secondary volcanic deposits 65

4.2 Plutonics . . . 66
4.21 Basement rocks . . . 66
4.22 Plutonic intrusions . . . 68
4.23 Plutonic extrusions . . . 68

4.3 Reduction and weathering . . . 70
4.31 Chemical reduction . . . 70
4.32 Physical reduction . . . 71
4.33 Contrition . . . 74
4.34 Corrasion by wind blown sand . . . 76
4.35 Biological agents . . . 78

4.4 Sedimentary rocks . . . 79
4.41 General aspects . . . 79
4.42 Sedimentary rocks in situ . . . 80
4.43 Sedimentary structures . . . 84

4.5 Metamorphics . . . 90
4.51 Metasediments . . . 90
4.52 Metavolcanics . . . 92

5 Slope development . . . 95

5.1 Slope types . . . 96
5.11 Normal slope recession . . . 96
5.12 Special erosional features . . . 98

5.2 Aqueous erosion . . . 101
5.21 Random patterns . . . 101
5.22 Badlands . . . 103
5.23 Unstable erosion . . . 104

5.3 Loose materials . . . 106
5.31 Scree slopes . . . 106
5.32 Scree cones . . . 111

5.4 Slow motion phenomena . . . 112
5.41 Creep and slumps . . . 112
5.42 Soil drag . . . 118
5.43 Tear-scars . . . 118
5.44 Ground fissures . . . 122
5.45 Grass slides . . . 122
5.46 Mountain fracture . . . 124

5.5 Fast movement on slopes . . . 127
5.51 Rock falls . . . 127
5.52 Land slides . . . 128

6 River action . . . 131

6.1 River bed processes . . . 132

6.11 River flow 132
6.12 Bed load 134
6.13 Movable bed instabilities 136
6.14 Rocky river bed 138

6.2 Vertical river action 140
6.21 Mountain torrents 140
6.22 Gorges and canyons 142
6.23 Waterfalls 147

6.3 Lateral river action 154
6.31 Bank erosion 154
6.32 Meanders 158
6.33 River terraces 162

6.4 Valleys 164
6.41 Upper reach 164
6.42 Middle reach 166
6.43 Lower reach 168
6.44 Tectonic design of valley trend 168

6.5 Solution and deposition effects of water 170
6.51 Solution effects 170
6.52 Karst 172
6.53 Deposition effects 176

7 Large bodies of water 179

7.1 Morphology 180
7.11 The sea surface 180
7.12 Ocean bottom and marine orogenesis 180
7.13 Eustatic changes 182
7.14 Islands 184

7.2 Waves and currents 184
7.21 Waves 184
7.22 Currents 188

7.3 Coasts 188
7.31 Steep coasts 188
7.32 Shallow coasts 193
7.33 Coral reefs 202
7.34 Shore erosion 206

7.4 River mouths 210
7.41 Estuary 210
7.42 Delta 210
7.43 Fyord 212

8 Niveal features 213

8.1 The snow/ice surface 214
8.11 Snow fields 214
8.12 Glacier morphology 216
8.13 Ice caps 222

8.2 Glacier action 224
8.21 Glacial erosion 224
8.22 Moraines 227
8.23 Erratic blocks 229

8.3 Other (peri-)glacial features 230
8.31 Patterned ground 230
8.32 Periglacial deposits 230
8.33 Glacial humps 234
8.34 Periglacial dead-ice 236

9 Desert features and related phenomena 239

9.1 Aeolian features 240
9.11 Clouds 240
9.12 Dunes 240
9.13 Aeolian ripples 244

9.2 Evaporite environment 245
9.21 Evaporite formation 245
9.22 Evaporite flats 248

9.3 Rocky deserts 249
9.31 General aspects 249
9.32 Desert rocks 250

10 Volcanic landscapes 253

10.1 Volcanic forms 254
10.11 Volcano shape 254
10.12 Volcanic crater 256
10.13 Volcanic plugs 260

10.2 Volcanic effects 262
10.21 Lava flow 262
10.22 Volcanic bombs 266
10.23 Lapilli 268
10.24 Lahar 270
10.25 Caldera 271

10.3 Hydrothermal effects 272
10.31 Geysers 272
10.32 Hot springs 274

2 Basic landscape types

We have seen in the Introduction that the most fundamental principle active in landscape genesis is the principle of antagonism. A fundamental division of landscapes is thus obtained by considering whether the activity level of the antagonistic processes is high (Figs. 2.11 to 2.13) medium (Figs. 2.21 to 2.23) or low (Figs. 2.31 to 2.33).

At each activity level, a further distinction is obtained by considering the prevailing exogenic agents: This may be water (in a humid climate), ice (in a glacial climate) or wind (mainly, but not exclusively, in an arid climate). Additional subdivisions may have to be introduced on account of the material underlying a landscape. In this fashion, the taxonomy illustrated by the pictures now following was arrived at.

Fig. 2.11 A. *High-activity landscape, humid climate: hard crystalline rocks.* High-level activity in a humid climate occurring in hard crystalline rocks is one of the most common instances of the operation of the principle of antagonism in geomorphology. In mountain ranges in low to medium geographical latitudes, such as in the Alps, Himalayas, Rocky Mountains or Andes, deeply cut valleys and rounded granitic peaks are produced. The picture shows the Urubamba Valley near Machu Picchu in Peru

Fig. 2.11 B. *High-activity landscape, humid climate: hard sedimentary rocks.* In sedimentary rocks under high-activity humid conditions, it is the peaks rather than the valleys which become abrupt and craggy. The picture shows the Mythen in the Canton of Schwyz in Switzerland, indicating these conditions: a calcareous massif that has been thrust from the south over softer, younger rock to its present position. Exogenic agents created its present form

Fig. 2.11A

Fig. 2.11B

Fig. 2.11C. *High-activity landscape, humid climate: soft rocks.* If the substratum is soft, more rounded forms evolve even under high-activity conditions. Such a soft substratum is represented by volcanic ash and quickly-weathering volcanic deposits. A typical landscape of this type is found in the volcanic highlands of the island of Bali, Indonesia

Fig. 2.12. *High-activity landscape, glacial climate.* In a cold climate, which may be found either in middle latitudes at high altitudes or else in high latitudes, it is the glaciers rather than the rivers which shape the landscape. Shown here is a high-activity glacial landscape near the Tierbergli above the Steingletscher in the Canton of Berne, Switzerland

Fig. 2.11C

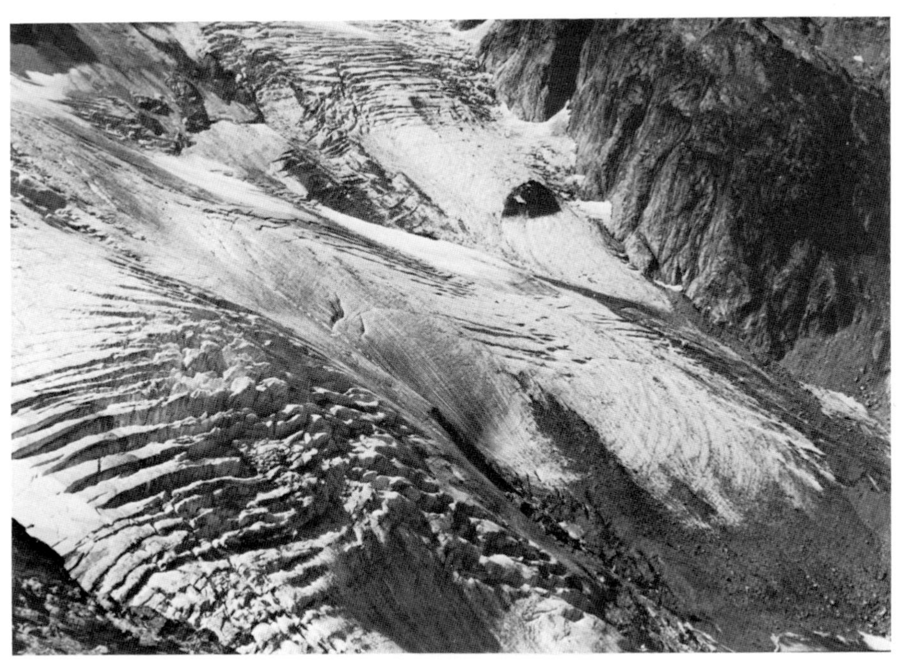

Fig. 2.12

Fig. 2.13. *High-activity landscape, arid climate.* In an arid climate, high activity leads to bare rocky desert slopes. The degradation is caused by occasional severe cloudbursts during which the dry valleys become filled with water. Such a desert exists on the southern slopes of the Rocky Mountains near Tucson in Arizona, USA

Fig. 2.21 A. *Medium-activity landscape, humid climate: hard metamorphic rocks.* The medium-activity level produces gentle features on a metamorphic substratum which may be seen in the Bohemian massif in Lower Austria. The picture was taken near the Czech border; the castle in the foreground is Hardegg castle

Fig. 2.13

Fig. 2.21 A

Fig. 2.21B

Fig. 2.21B. *Medium-activity landscape, humid climate: sedimentary rocks.* The rounded hills of the Appalachians (near Clingman's Dome, Tennessee, USA) are typical of a sedimentary, humid, medium-activity landscape

Fig. 2.22

Fig. 2.22. *Medium-activity landscape, glacial climate.* The effect of a glacial climate becomes evident after the glaciers have disappeared. This occurred after the last ice age on Baffin Island, Canada where the relief corresponds to a medium-activity landscape

Fig. 2.23. *Medium-activity landscape, arid climate.* In a dry climate under medium-activity conditions, the wind causes the most important geomorphological features by shifting sand or silt and clay to form dunes. The picture above shows the "Chinese wall": silt-clay dunes windswept from an old lake bed during a dry phase (dry Lake Mungo area, N.S.W., Australia)

Fig. 2.31 A. *Low-activity landscape, humid climate: hard substratum.* On a crystalline substratum in a humid climate the base level of erosion ("low activity") is reached with some buttes and mesas being left (instability principle in erosion). The picture shows a scene illustrating such conditions from the Deccan traps in India, near Poona (after Scheidegger and Padale, 1982; also Scheidegger, 1982)

Fig. 2.23

Fig. 2.31A

Fig. 2.31 B. *Low-activity landscape, humid climate: soft substratum.* On a soft substratum, no mesas can be formed and, at the low-activity level, and alluvial plain may develop, as shown above: scene in central Alberta, Canada, in winter

Fig. 2.32. *Low-activity landscape, glacial climate.* The scouring action of ice that was formerly present, is particularly easily recognizable in high northern latitudes where all vegetation is absent (Barren Lands in the Canadian Northwest Territories, Keewatin District). Before the melting of the ice occurred after the end of the last ice age, the base level of erosion had been reached

Fig. 2.31B

Fig. 2.32

Fig. 2.33 A. *Low-activity landscape, arid climate: soft substratum.* Low-level activity leads in arid climates to a desert plain; in soft materials, this is a sandy desert as shown in the above picture taken in the Algerian Sahara

Fig. 2.33 B. *Low-activity landscape, arid climate: hard substratum.* On a hard substratum, a low-level activity desert develops into a stony plain, as represented here by the air-strip near Ayers Rock in Central Australia

Fig. 2.33A

Fig. 2.33B

3 Structural background

As noted in the Introduction, the structural (tectonic) background plays a fundamental role in landscape genesis, particularly with regard to the predesign of features later affected by exogenic agents. This background is the consequence of endogenetic processes which produce stresses in the rocks forming the surface of the earth. The response of the rocks to these stresses is by undergoing deformations. These may either be continuous (Figs. 3.11 to 3.14) or discontinuous (Figs. 3.21 to 3.23).

In addition, the geotectonic stresses produce specific effects on surfaces (wall effects and exfoliations; Figs. 3.31 and 3.32); furthermore, in connection with the selection principle mentioned in the Introduction, characteristic degradational features are created due to self-gravitational effects (Figs. 3.41 and 3.42).

Fig. 3.11 A. *Tectonic fold: anticline.* Tectonic folds are either domed upwards ("anticline"), downward ("syncline"), or they are a combination of the two possibilities joining one level with a higher or lower one ("monocline"). Shown here is a small anticline in an outcrop on a lake in the Grenville Province of the Canadian Shield (Ufford, Ontario, Canada) (after Scheidegger, 1976)

Fig. 3.11 B. *Tectonic fold: syncline.* An example of a small syncline was found in Triassic dolomites in the Hochkönig area of Salzburg Province, Austria

Fig. 3.11A

Fig. 3.11B

Fig. 3.11C. *Tectonic fold: monocline.* Some "mountains" are the result of monoclinal folding. The Blue Mountains of New South Wales (Australia) are of this type. As seen in the photograph, the strata are almost perfectly flat and yet, the hills have the superficial appearance of low mountains, owing to river erosion. The monoclinal folding seems to have occurred in the Pliocene time (scene near Katoomba in New South Wales)

Fig. 3.11D. *Tectonic fold: small-scale.* Small-scale tectonic folds may be seen in the outcrop near Ste. Flavie, Québec, Canada, shown above: lower Paleozoic terrigenous sediments have been folded during the Taconian (Ordovician) orogeny (Appalachian system); the scale is rather small as evidenced by the coin which is a Canadian quarter

Fig. 3.11C

Fig. 3.11D

Fig. 3.12A. *Folds from differential movement: flow-folds.* A special type of "folds" are flow folds which are caused by the differential movement of laminae normal to the layering: in this fashion, the original layering is distorted and presents the appearance of folds (Precambrian migmatite near Parry Sound, Ontario, Canada)

Fig. 3.12B. *Folds from differential movement: chevron folds.* Another type of folds resulting from differenital movements are "chevron folds": these result from movements corresponding to those in a stack of cards. The chevron folds shown here were found in Paleozoic shale in the Babo Shan fault zone near Beijing, China

Fig. 3.12A

Fig. 3.12B

Fig. 3.13A. *Instability phenomena: boudinage.* Boudinage occurs if a competent (hard) layer is squashed between incompetent (soft) ones: the hard layer is broken into sausage (boudin-)like structures as evidenced by the amphibolite layer in marble shown here (Bohemian massif, Unterthürnau, Lower Austria)

Fig. 3.13B. *Instability phenomena: pinch-and-swell structures.* If the harder layer squashed between softer ones does not actually break, one speaks of pinch-and-swell structures as seen here in a gneiss near Ronco, Canton of Ticino, Switzerland

Fig. 3.13A

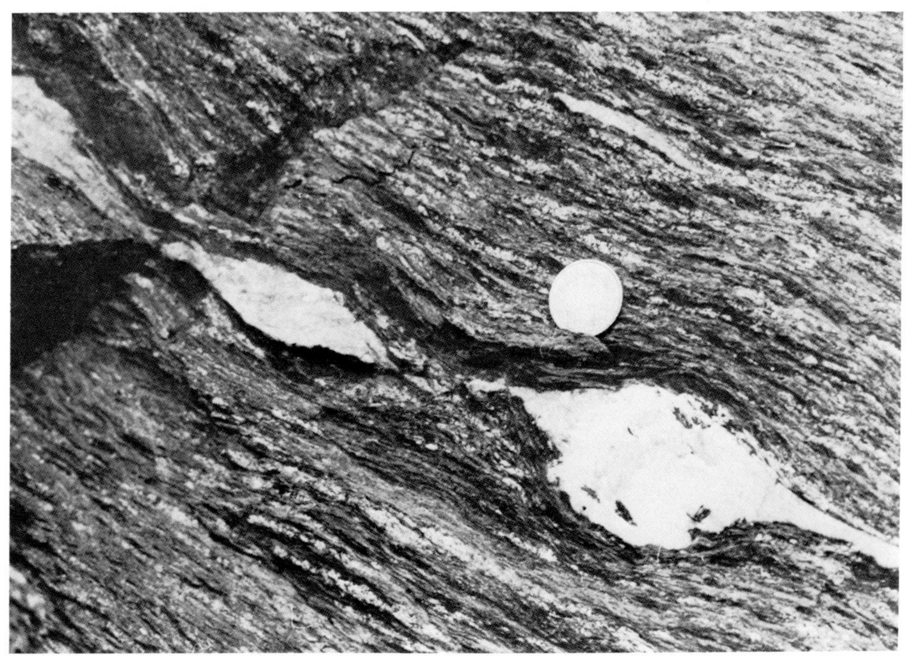

Fig. 3.13B

Fig. 3.14A

Fig. 3.14A. *Deformation of petrofabrics: squashed fossils.* The stressing of the rocks may be evident in the deformation of fossils, as in the stretched belemnite in Lias slate near Martigny, Canton of Valais, Switzerland, shown here

Fig. 3.14B

Fig. 3.14B. *Deformation of petrofabrics: squashed reduction spots.* Reduction spots, e.g. in Permian mudplain deposits near Quarten, Canton of St. Gall, Switzerland, started out as spherical features. Upon being stressed by tectonism they became squashed into ellipsoids

Fig. 3.21 A. *Joints: subparallel sets.* Joints are small fractures in rock which have been caused by tectonic stresses. They are seen everywhere in outcrops. Usually, they do not occur at random, but form distinct subparallel sets. The picture shows such a set in lower Paleozoic terrigenous sediments that have been affected by the Taconian (lower Ordovician) orogeny near St. Jean-Port-Joli on the banks of the St. Lawrence River in Québec, Canada

Fig. 3.21 B. *Joints: conjugate systems.* The joint sets are usually 2 or 3 in number (one horizontal, two vertical) and intersect each other at angles which are not quite 90°. The picture shows two such (vertical) sets in granite near Hyderabad, India (after Scheidegger and Padale, 1982)

Fig. 3.21A

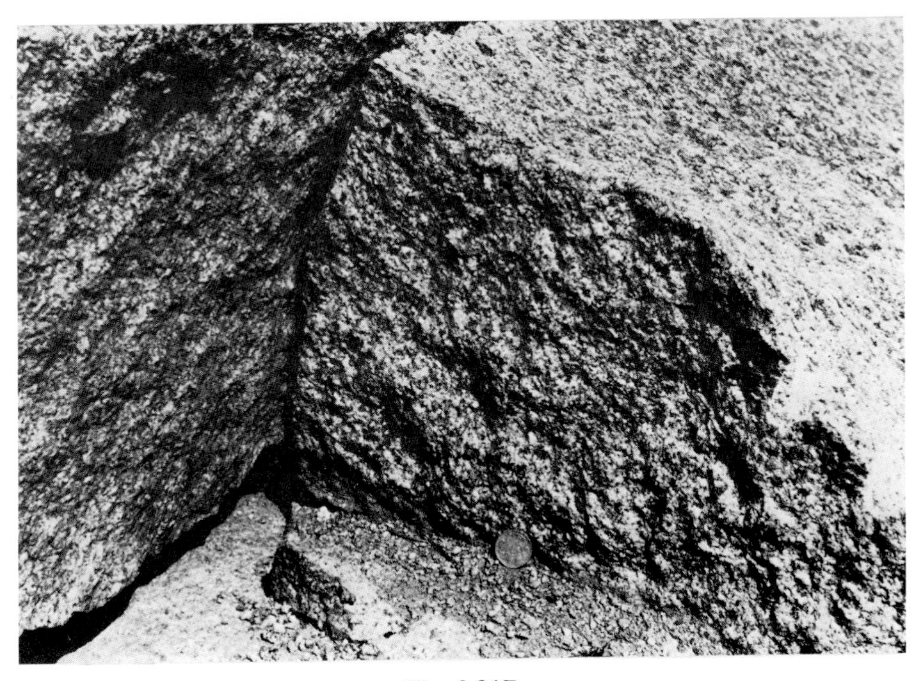

Fig. 3.21B

Fig. 3.21 C

Fig. 3.21 C. *Joints: parallelepiped.* The two conjugate joint-sets in an area commonly give rise to the formation of characteristic parallelepipeds. Shown here is the fundamental parallelepiped in a hornblende-diopside-gneiss outcrop in the Grenville province of the Canadian Shield, near Ufford, Ontario, Canada (after Kohlbeck and Scheidegger, 1977)

Fig. 3.21 D

Fig. 3.21 D. *Joints: niche.* The mirror image of a joint-parallelepiped is a niche. The one shown here was found in a basalt lava flow on Tenerife (after Scheidegger, 1978)

Fig. 3.22 A

Fig. 3.22 A. *Faults: in rocks.* Faults are simply large fractures in rocks. Like joints, they manifest themselves in subparallel sets, but on a much larger scale; as seen here on the Zuetribi Stock, in the Canton of Glarus, Switzerland (Malm-limestone)

Fig. 3.22B

Fig. 3.22B. *Faults: on mountain side.* On a mountain side, faults manifest themselves in the predesign of the gullies. Shown here is the flank of Piz Ot in the Samnaun valley, Canton of Grisons, Switzerland

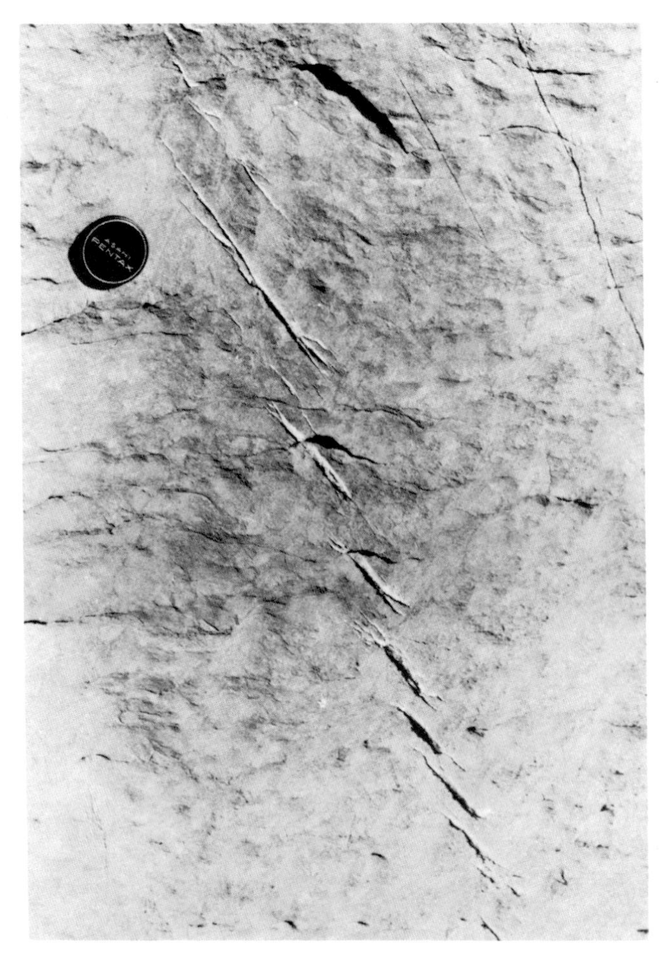

Fig. 3.22C

Fig. 3.22C. *Faults: en echelon.* Faults occur often "en echelon", particularly in shear zones, as seen here in Valenginian (Cretaceous) limestone at the Col de Morcle, Canton of Valais, Switzerland. The cracks have later been filled by calcite precipitating out of formation water

Fig. 3.23A

Fig. 3.23A. *Thrust sheets: hog's backs.* The great land-based mountain ranges of the Earth are generally the consequence of a "crustal shortening" (due to the collision of tectonic plates) of continental dimensions. The shortening causes the formation of thrust sheets which are pushed over each other. Thus, as one approaches the Rocky Mountains from the prairies, one finds that higher and higher thrusts follow each other; one of these presents a weird sequence of "hog's backs" near Colorado Springs, Colorado, USA

Fig. 3.23B. *Thrust sheets: mountain range.* In the Rocky mountains proper, the thrusts become very large indeed. Thus many mountains have a relatively flat "back" and a very steep "front". The back represents the plane of the thrust sheet, the front is where the sheet may give rise to a long chain of mountain peaks according to the manner in which the break-off at the front occurred. In this fashion a most spectacular landscape was created as exemplified by the Endless Chain Ridge in Alberta, Canada, shown here

Fig. 3.23C. *Thrust sheet: ledge.* The ledge where the "back" and the "front" of a mountain meet, may be very narrow and quite a challenge to a mountaineer. The ledge shown here is that at the summit of Mount Rundle, Alberta, Canada

Fig. 3.23B

Fig. 3.23C

Fig. 3.31A

Fig. 3.31 A. *Wall effects: formation of a couloir.* On a wall the stresses are such that vertical conjugate shearing surfaces are formed which give rise to a "couloir", similar to a "niche" in a joint system. Shown here is a couloir formed in a calcareous wall on the Zeiher Homberg in the Swiss Jura mountains. The "roof" of the couloir is the result of the diminishing vertical stresses as the top of the wall is reached

Fig. 3.31 B

Fig. 3.31 B. *Wall effects: saw-toothed ledge*. On a protruding ledge, the conjugate joints can produce a saw-toothed sequence of fractures instead of a single couloir (Gisliflue, Swiss Jura Mountains)

Fig. 3.31 C

Fig. 3.31 C. *Wall effects: crags on a plateau.* In the plateau on top of a wall, the wall fractures produce conjugate crags that may be leached out to a treacherous width. Shown are the crags in a plateau of Silurian dolostones forming the Niagara escarpment, near Mt. Nemo, Ontario, Canada

Fig. 3.31 D

Fig. 3.31 D. *Wall effects: recession.* A wall recedes from the bottom up, otherwise it would rapidly become inclined and end up as gentle slope bank. At the foot of a wall, stress concentrations occur which favor the decay: breakouts cause a notch to be formed; the wall above it comes off in sheets and thus the wall remains vertical while it recedes (wall of the Niagara canyon, Niagara Falls, Ontario, Canada)

Fig. 3.32 A

Fig. 3.32 A. *Exfoliation: gravitational effect.* As noted, the gravitational stresses in a wall cause exfoliations, in which whole sheets of material are split off parallel to the wall. This phenomenon is seen here in the Jurassic limestone of the Strihen, Canton of Aargau, Switzerland

Fig. 3.32B

Fig. 3.32B. *Exfoliation: induced by weathering*. Exfoliation can also be induced by weathering effects, inasmuch as the top layer of a granite is affected by the water penetrating into it; thus it will peel off (exfoliation in granite near Granite City, Missouri, USA)

Fig. 3.41 A

Fig. 3.41 A. *Self-gravitational effects in mountains: single peaks.* The selection principle of geomorphology states that degradation occurs in such a fashion that statically stable forms are preferentially "selected": In the Matterhorn (Canton of Valais, Switzerland) the ledges can be considered as support-struts which hold the structure up. The degradation in this case was caused mainly by ice action

Fig. 3.41 B

Fig. 3.41 B. *Self-gravitational effects in mountains: towers and bastions.* On a mountain ledge, the statically most stable forms are single bastions or towers that have the shape of the Eiffel tower. Shown here are the "Three Sisters", near Katoomba, N.S.W., Australia (after Gerber and Scheidegger, 1973)

Fig. 3.42

Fig. 3.42. *Self-gravitational effects on plateau edge.* On the edge of a plateau, the statically most stable forms selected by the erosion are garlands of heads: this is seen for instance on the edge of a Jurassic plateau near Effingen (Canton of Aargau, Switzerland) (after Scheidegger, 1985 a)

4 Petrological background

In addition to tectonics, the petrology also affects the morphology of a landscape. Basically, one distinguishes between two types of rocks: igneous rocks and sedimentary rocks.

The igneous rocks are presumably the "primary" ones: they have solidified from melts issuing from the interior of the earth, either after being discharged from volcanoes as "volcanics" (Figs. 4.11 to 4.13), or by rising, intruding and crystallizing at depth as "plutonics" (Figs. 4.21 to 4.23).

The rocks on the surface of the earth are then subject to reduction and weathering (Figs. 4.31 to 4.35), they are transported away and redeposited by exogenic agents as "sediments" (Figs. 4.41 to 4.43). Eventually, they may be subject to high pressures and temperatures so that they become metamorphosed. The "metamorphic" rocks may present an aspect close to that of plutonic rocks (Figs. 4.51 and 4.52).

Fig. 4.11 A. *Volcanic flows: andesite.* The melts issuing from volcanoes are called "lavas". Lavas are either andesitic (high in silica content) or basaltic (low in silica content). Andesitic lavas are found mainly on continents and continental plate boundaries, basaltic lavas in oceanic regions. Andesite crystallizes as a coarse-grained material in a lava flow, as seen here on the Dieng Plateau in central Java, Indonesia

Fig. 4.11 B. *Volcanic flows: ropy lava.* Andesitic lava is very viscous and thus forms a "ropy" structure upon cooling. Shown here is the ropy lava in an outcrop near the National University of Mexico in Ciudad de México

Fig. 4.11 A

Fig. 4.11 B

Fig. 4.11 C. *Volcanic flows: basaltic lava.* If the lava is basaltic, it often consolidates in form of polygonal columns which, when weathered and eroded, form polygons on the ground (picture taken near Langtang, Plateau State, Nigeria)

Fig. 4.11 D. *Volcanic flows: ripples.* On occasion, a lava flow forms ripples which become preserved upon cooling, as seen here in Nigeria, near Gimi Junction (Plateau State, Nigeria)

Fig. 4.11C

Fig. 4.11D

Fig. 4.11 E. *Volcanic flows: diabase-spilite pillow lavas.* Lava flows that enter the sea solidify in the form of "pillows". Shown here is a diabase-spilite pillow lava in the Ebriach gorge (Carinthia, Austria) on the Priadriatic lineament (after Kohlbeck et al. 1980)

Fig. 4.12 A. *Volcanic ejecta: ignimbrites.* Not all rock materials issuing from a volcanic crater do so as liquid melts: Some appear as "ejecta", like the ignimbrites which have been formed from a hot, turbulent, rapidly expanding magmatic gas. They are often welded to form banked deposits as the ignimbrites shown here near the Hilton Hotel in Addis Ababa, Ethiopia

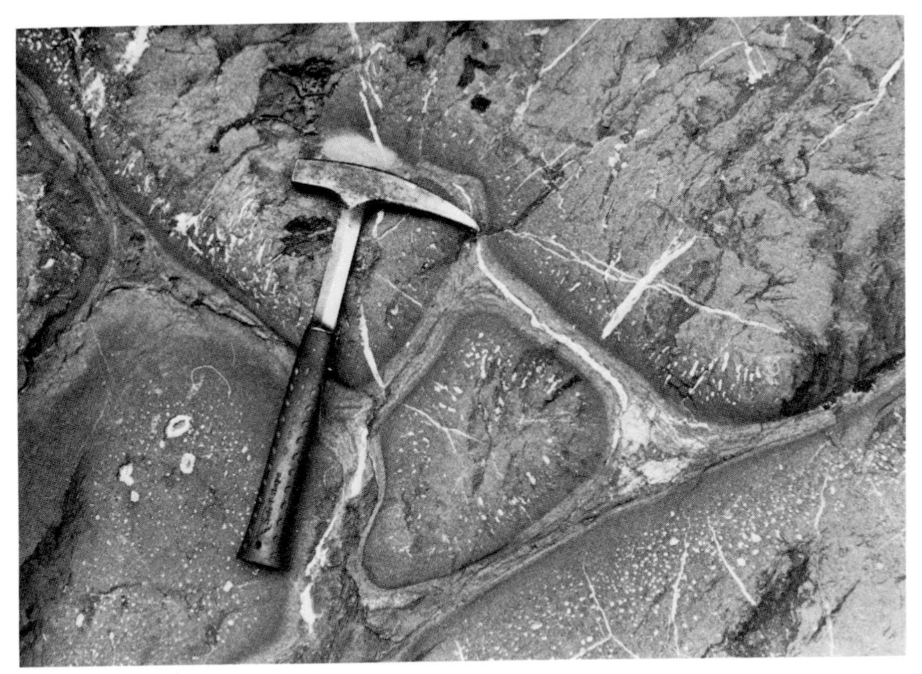

Fig. 4.11E

Fig. 4.12A

Fig. 4.12B

Fig. 4.12B. *Volcanic ejecta: tuff.* If the volcanic ejecta are loose and only slightly welded, they are called "tuffs". Accordingly, they are quite prone to being eroded, as demonstrated by this rain-washed volcanic tuff on the Dieng Plateau in central Java, Indonesia

Fig. 4.13A

Fig. 4.13A. *Secondary volcanic deposits: lahar.* Secondary volcanic deposits are created if the primary deposits are moved by water or wind to a new location. Some of the most important such deposits are lahars: the result of a water-borne debris flow that may be found at some distance from the volcanic center. Shown here is a lahar deposit on the island of Guadeloupe (Basse Terre, near Bouillante), F.W.I.

Fig. 4.13B. *Secondary volcanic deposits: fluvio-volcanics.* If the volcanic deposits are redeposited by rivers, one speaks of "fluvio-volcanics". Shown here is a deposit of remnants of heavily weathered and laterized basalts on the Jos Plateau, Plateau State, Nigeria

Fig. 4.21. *Basement rocks.* Basement rocks are seen in the root zones of the great mountain ranges of the world. Shown here is the basement granite in the Indus root zone of the Himalaya near Thikse, Ladakh, India

Fig. 4.12B

Fig. 4.21

Fig. 4.22. *Plutonic intrusion.* Plutons may be intruded into the surrounding country rock. Later, if they become exposed, they form "inselbergs" (island mountains). One of the largest such bergs is Stone Mountain, 24 km NE of Atlanta, Georgia, USA), which rises as a granite dome some 500 m out of the Appalachian Piedmont plain

Fig. 4.23. *Plutonic extrusion.* Extrusions can lead to a topographic relief in the manner exemplified by Mount Wellington, overlooking the city of Hobart in Tasmania (Australia). There was here a Devonian orogeny, but it was completely peneplained. In the Jurassic time, dolerite was extruded over the plain, and in the Tertiary epoch block faulting occurred which produced the relief

Fig. 4.22

Fig. 4.23

Fig. 4.31

Fig. 4.31. *Chemical reduction.* The reduction of rocks occurs by many processes. Most people probably think first of chemical agents, but these are effective only under certain very specific conditions. An example of the chemical actions of sea water on intertidal flats (i.e. flats lying below high tide level and above low tide level) is shown here (Fécamp, Normandy, France)

Fig. 4.32A

Fig. 4.32A. *Physical reduction: on sea shore.* Physical agents are much more effective in the reduction of rocks than chemical ones. Particularly if the rock is already cracked, the weathering elements are able to penetrate into the cracks; freezing and thawing may take effect and break the rock into small pieces. Cliff at Fécamp, Normandy, France

Fig. 4.32B

Fig. 4.32B. *Physical reduction: effect of layering*. Different layers are affected differently by the weathering agents. In the Milagro formation, near Maracaibo (Est. Zulia), Venezuela, layers of (Quaternary) lake sediments alternate with laterite crusts that were formed when the lake level was low. The weathering produced a startling effect of a sculptured wall

Fig. 4.32C

Fig. 4.32C. *Physical reduction: effect of differential resistance.* Limestone concretions in a sandstone produce zones of differential resistance to weathering. The resulting features have been called "knauer". (Molasse sandstone in a valley near Meilen, Canton of Zürich, Switzerland)

Fig. 4.32D. *Physical reduction: tectonic predesign.* A joint system may serve as a predesigning factor for the attack of the exogenic agents. This is seen particularly in the granulites of the region around Dürnstein in Lower Austria (Bohemian massif)

Fig. 4.33A. *Contrition: on coast.* The mechanical action of water in the surf on a coast may reduce the size of the loosened rocks to that of pebbles. This process is called contrition. The result of it is seen on the beach below the cliffs of Fécamp, Normandy, France

Fig. 4.32D

Fig. 4.33A

Fig. 4.33 B. *Contrition: in a river bed.* Contrition of rocks also occurs in river beds by the continual rolling and dragging caused by the streaming water. In a few tens of kilometers the bed load pieces become quite rounded even in a mountain stream (Lech river, Tyrol, Austria, near Dürnau)

Fig. 4.34. *Corrasion by wind-blown sand.* Mechanical reduction of rocks may be caused by the incessant attack of some external elements. This process is called "corrasion". A particularly interesting case is aeolian corrasion, in which blown sand may produce rocks that appear almost polished. The specimen shown here was found in the Algerian Sahara

Fig. 4.33B

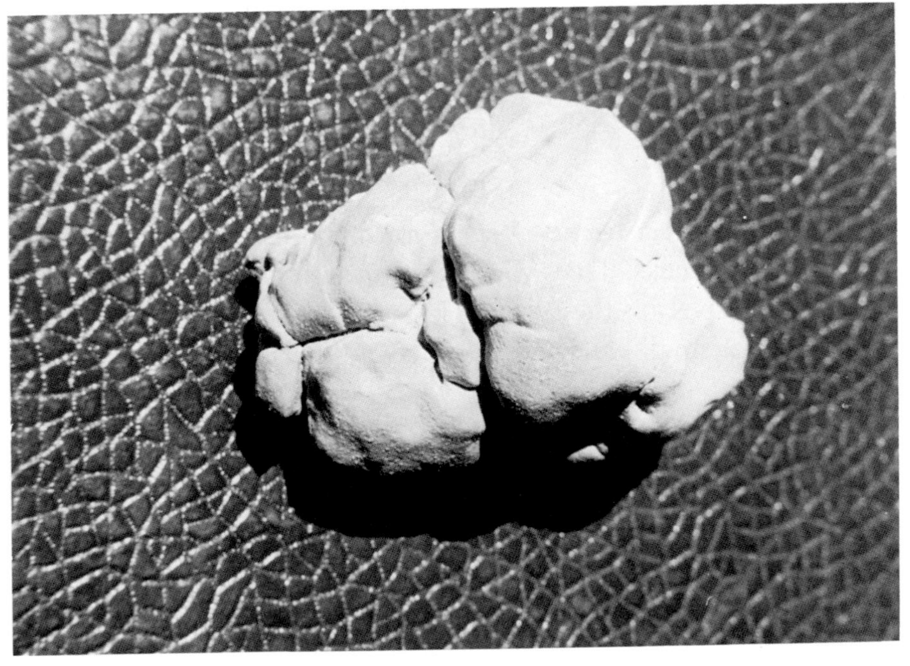

Fig. 4.34

Fig. 4.35

Fig. 4.35. *Biological agents.* Vegetation may also contribute to the destruction of rocks. Trees will take root in any crack, forcing it to open progressively as they grow. The scene shown here was recorded in a small valley near Exshaw, Alberta, Canada

Fig. 4.41

Fig. 4.41. *General aspects of sedimentary rocks.* The characteristic feature of sedimentary rocks is that they are layered corresponding to the deposition by various agents. Shown here is a layered sequence of clay, silt, sands and gravels of marine origin intercalated with lahar and tuff deposits in the Kalibang formation (Plio-Pleistocene) in the Sangiran Dome, Central Java, Indonesia

Fig. 4.42A. *Sedimentary rocks in situ: fossils.* During deposition, animals may become entrapped in the deposits as seen in the Plio-Pleistocene clay near Sangiran, Central Java, Indonesia

Fig. 4.42B. *Sedimentary rocks in situ: molasse.* Large debris-pieces from a rising mountain range form a conglomerrate at the foot of the range which is called "molasse". Shown here is the conglomerate ("Nagelfluh") of the upper fresh water molasse (Tertiary) on the flank of the Atzmännig, Canton of St. Gall, Switzerland

Fig. 4.42A

Fig. 4.42B

Fig. 4.42C

Fig. 4.42C. *Sedimentary rocks in situ: sandy shale.* Sequences of sands and clay lead, after the occurrence of orogenic stresses, to the creation of sandy shales as seen in the Werfen layers at the base of the Triassic in the Hochkönig Massif, Salzburg Province, Austria

Fig. 4.42D

Fig. 4.42D. *Sedimentary rocks in situ: evaporites.* Evaporite forms bands in surrounding clay as shown here in the mine wall of the Bad Ischl salt mine: The halite shows up as a dark (actually: red) band in the surrounding Permian shale (Upper Austria)

Fig. 4.43 A. *Sedimentary structures: runing-water laid cross beds.* If the deposition of sediments occurs in running water, the local layering may not be parallel to the main sedimentation plane: wandering dunes in a stream produce "cross-beds". An example is shown here from a Pleistocene glacial out-flow deposit near Ufford, Ontario, Canada

Fig. 4.43 B. *Sedimentary structures: deep sea cross beds.* In deep-sea deposits, cross beds are caused by slumping and turbidity currents. This is the origin of the cross beds in Mesozoic schists (Bündnerschiefer) in the Klus near Landquart, Canton of Grisons, Switzerland

Fig. 4.43A

Fig. 4.43B

Fig. 4.43 C. *Sedimentary structures: fluting.* Deep-sea deposits do not only show cross-bedding, but also flutings: these are casts of the channel rills caused by deep-sea currents. The fluting shown here was found in Cretaceous Bündner-schist (deposited in the Penninic trough) outcropping in the Klus above Landquart, Canton of Grisons, Switzerland

Fig. 4.43 D. *Sedimentary structures: sole marks.* Flysch was deposited in deep water: slow settling of clay was interrupted by the occurrence of turbidity currents. During the quiet phase of clay deposition, ocean currents produced scour holes with a sharp, semicircular edge pointing up-current. These were filled by the sand from turbidity currents; after consolidation and tectonic displacement the resulting sandstone shows a cast of the holes on its underside: the sole marks. This is particularly evident in the flysch of the Vienna Woods in Austria

Fig. 4.43C

Fig. 4.43D

Fig. 4.43E

Fig. 4.43E. *Sedimentary structures: ball-and-pillow structures*. The strange ball-and-pillow structures in the flysch of the Vienna Woods (Austria) are the result of the foundering of sand layers into underlying clay which was caused by the abrupt thixotropic liquefaction of the latter

Fig. 4.43F

Fig. 4.43F. *Sedimentary structures: truncation.* Cross-bedded sand layers may be further eroded at the top; this produces a truncation above which newer sediments are subsequently deposited. The example shown is from the Edmonton sandstone (Cretaceous) near Drumheller, Alberta, Canada

Fig. 4.43G. *Sedimentary structures: ice-laid deposits.* In contrast to water-laid structures, ice-laid structures have a "jumbled" appearance like the conchoidal feature shown here which is the result of the melting of an ice lens in sand after the end of the last ice age (Muskoka region, Ontario, Canada)

Fig. 4.51A. *Metasediments: slightly metamorphosed.* The transition between sedimentary and metamorphic rocks is continuous. Shown here are lower Paleozoic terrigenous sediments affected by the Taconian and Acadian orogenies. Lower St. Lawrence region, near St. Jean-Port-Joli, Québec, Canada

Fig. 4.43G

Fig. 4.51A

Fig. 4.51 B. *Metasediments: complete anatexis.* The Grenville rocks in northern Ontario consist mainly of banded "migmatites" which indicate the previous sedimentary layering. However, when complete anatexis (melting) and recrystallization has occurred, garnets may grow in the melt, as seen in an outcrop near Parry Sound, Ontario, Canada

Fig. 4.52. *Metavolcanics.* When the metamorphism affects volcanic flow, the original block type structure may be preserved as in an outcrop of metavolcanics near Bancroft, Ontario, Canada

Fig. 4.51B

Fig. 4.52

5 Slope development

The various landscapes of the world are made up of *slopes*: mountain sides, river beds, and crest lines are all examples of slopes. If it can be understood how slopes change under the influence of external agents, then it is obviously possible to explain physical geography.

In a relief that has been caused or is being maintained by endogenic processes, slopes are subject to recession. The recession leads to some standard or rather special forms (Figs. 5.11 and 5.12).

The process of recession is commonly caused by water flowing across a slope. It may result in some rather peculiar and characteristic features (Figs. 5.21 to 5.23). On occasion, the slope recession also occurs through the slow spontaneous movement in loose (Figs. 5.31 and 5.32) or compact (Figs. 5.41 to 5.46) materials; in rare circumstances, such movements may become catastrophically rapid (Figs. 5.51 and 5.52).

Fig. 5.11 A. *Normal slope recession: linear.* The various denudational agents combine to produce slope evolution on a large scale. Some of the slopes simply seem to recede, presenting at all times a perfectly straight profile (linear slope recession: this case has been postulated as the "normal" one by Penck, 1929). The above picture shows a typical case: A mountain side near Einsiedeln, Canton of Schwyz, Switzerland

Fig. 5.11 B. *Normal slope recession: S-shape.* A mathematical calculation (Scheidegger, 1970) yields that straight slope recession, as postulated by Penck (1929) is, in fact, an exceptional case. A denudational slope will generally evolve into an S-shaped profile with a broad, flat toe and a somewhat sharper bend at the top. An example from a molasse hill near Gunterswil, Canton of Lucerne, Switzerland, is shown here

Fig. 5.11A

Fig. 5.11B

Fig. 5.12A

Fig. 5.12A. *Special erosional features: earth pillars.* Occasionally, denudation on slopes can produce interesting features. Here, in the desert near Lamayuru (Ladakh, India), a series of pillars have been left standing whilst the slope was eroded underneath

Fig. 5.12B

Fig. 5.12B. *Special erosional features: hoodoos.* Sometimes, if some layers are more resistant to erosion than others, very weird structures may be formed. The hoodoo in the Drumheller Badlands (Alberta, Canada) shown here has an overhanging hat. Presumably, the neck of the hoodoo was eroded by rain-water that turned around the brim of the hat just like tea running down the underside of the spout of the tea pot (Scheidegger, 1958)

Fig. 5.12C

Fig. 5.12C. *Special erosional features: cap-rock mesa.* More conventional erosional features than those shown in the previous pictures are mesas; these are pyramidal structures where the top is formed by a somewhat more resistant layer than the clay below. (Photograph taken near Denver, Colorado, USA)

Fig. 5.21A

Fig. 5.21A. *Random patterns: general view.* The process of slope recession is a combined effect caused by various agents. The primary influence is that of water: little miniature gullies form which become enlarged as time passes, thereby causing destruction of the slope. The combinatorial structure of the junctions of the gullies can be shown to be *random* (Scheidegger, 1970). The picture was taken in a heap of mine tailings in Kickapoo Park, near Danville, Illinois, USA

Fig. 5.21B

Fig. 5.21 B. *Random patterns: single rill.* When a rill has been formed in a slope, it tends to accentuate itself, as shown here on a calcareous slope on the Isla de Toas, Est. Zulia, Venezuela

Fig. 5.22

Fig. 5.22. *Badlands*. The miniature gullying process is particularly impressive on clayey slopes with a relatively sparse cover of vegetation ("badlands"). (Picture taken in the Dry Lake Mungo area, N.S.W., Australia)

Fig. 5.23 A. *Unstable erosion: soft rock.* Eventually, the gullies with which the erosional and denudational processes begin, become deeper and broader. Soon they run into each other and form erosion funnels. This is also in conformity with the "instability principle" of geomorphology: a deviation form uniformity, e.g. a hollow, becomes bigger. Thus, catchment areas of a stream tend to widen out into a "cirque". Shown here is the head of a cirque on a molasse hill (Gurten near Berne, Switzerland)

Fig. 5.23 B. *Unstable erosion: hard rock.* An erosion cirque can even be formed in hard rock, as is demonstrated by the large cirque that developed in the basaltic Deccan traps near Ajanta shown in this picture (after Scheidegger and Padale, 1982)

Fig. 5.23A

Fig. 5.23B

Fig. 5.31 A

Fig. 5.31 A. *Scree slopes: General aspect.* A form of spontaneous mass movement occurs on scree slopes that may be found in high mountain country. Such scree slopes are simply accumulations of debris that are the result of the reduction of the rocks above. The scree on such slopes moves steadily downhill; this occurs in the form of miniature slips on account of the slope becoming overloaded with falling debris: the angle of the slope has a definite value which is that at which the rubble can just support itself. This angle is called the *angle of repose*. If this angle is exceeded because of the accumulation of debris falling down from the wall above, the slope becomes unstable and begins to slip (picture was taken near Kargil in the Suru Valley, Ladakh, India)

Fig. 5.31 B

Fig. 5.31 B. *Scree slopes: head section.* A scree slope normally develops at the foot of a wall. The scree accumulates from the debris falling down from above. Shown here is the head section of the scree slope near Hinterhard (Canton of Aargau, Switzerland) in the Jura mountains

Fig. 5.31 C. *Scree slopes: scree in middle section.* In the middle section, the surface of a scree slope shows a sieving effect: smaller rocks lie underneath bigger ones since they fall through the "sieve" formed by the latter. Schneeklamm, Hochkönig (Salzburg), Austria (after Gerber and Scheidegger, 1974)

Fig. 5.31 D. *Scree slopes: blocks on middle section.* Flat and prismatical blocks cannot roll or slide rapidly down a slope but are slowly pushed downhill by the scree accumulating behind them. This causes a rotation so that the lower part of large blocks protrudes from the slope. (Hochkönig area, Salzburg, Austria) (after Gerber and Scheidegger, 1974)

Fig. 5.31C

Fig. 5.31D

Fig. 5.31 E

Fig. 5.31 E. *Scree slope: end section.* At the bottom of a scree slope one finds generally large boulders which have been able to roll to this distance. The finer material was stopped higher up since the slope angle decreases with distance. The picture was taken on a scree slope near the Fuorcla Surlej, Canton of Grisons, Switzerland

Fig. 5.32

Fig. 5.32. *Scree cone*. If the material comes out of a narrow source, the scree slope assumes the form of a cone. Shown here is the scree cone in the lower Ochsenkar in the Hochkönig area (Salzburg) in Austria (after Gerber and Scheidegger, 1974)

Fig. 5.41 A

Fig. 5.41 A. *Creep and slumps: morphology of slope*. The morphology of a creeping slope is characterized by humps and hollows which give it a wavy appearance. The road from Sattel to Schwyz (Switzerland) is constantly being menaced by such a slowly creeping slope in flysch layers

Fig. 5.41 B

Fig. 5.41 B. *Creep and slumps: morphology of trees.* Mass movements on a slope are immediately indicated by the bent shape of the trees: As the slope creeps downhill, the roots of the trees stay fixed and the trees become inclined. Then, as the trees try to straighten themselves into the vertical, they grow in the form of a sabre or sickle. The region shown is in the flysch area of the northern Vienna Woods (Austria)

Fig. 5.41C. *Creep and slumps: destructive effects.* The mass movements have caused a road in the Caracas Valley (Venezuela) to become impassable

Fig. 5.41D. *Creep and slumps: large scale mass movement.* When the slow mass movements are of the order of meters per day and engulf a whole creek bed, the appearance of a slowly creeping worm is presented. The damage caused can be considerable (Bad Goisern, Upper Austria, March 1982)

Fig. 5.41C

Fig. 5.41D

Fig. 5.41 E

Fig. 5.41 E. *Creep and slumps: in clay*. A creeping clay slide shows a stepped appearance in conformity with the instability principle of geomorphology: this corresponds to the riffles, bars and pools in a river or brook (cf. Figs. 6.13). Shown here is a slide in opalinus clay (Jurassic) near Schinznach, Canton of Aargau, Switzerland: Note the stepped appearance

Fig. 5.41 F

Fig. 5.41 F. *Creep and slumps: instability effect.* Water collects in the hollows on a creeping slope, further destabilizing the latter. The picture was taken on the Kollmannsegg near Dienten, Salzburg, Austria (after Scheidegger, 1986)

Fig. 5.42. *Soil drag.* The mass movements in the surficial layers have frequently the effect that the layers beneath become bent, particularly if the latter consist of soft materials, as seen in the flysch zones of the Vienna Woods (Halterbach Valley, Vienna, Austria)

Fig. 5.43A. *Tear scar: general view.* Mass movements on a slope usually begin with a tear scar: a region bare of vegetation on which the motion is initiated. Shown here is the tear scar at the head of a large slide near Wörschach, Styria, Austria

Fig. 5.42

Fig. 5.43A

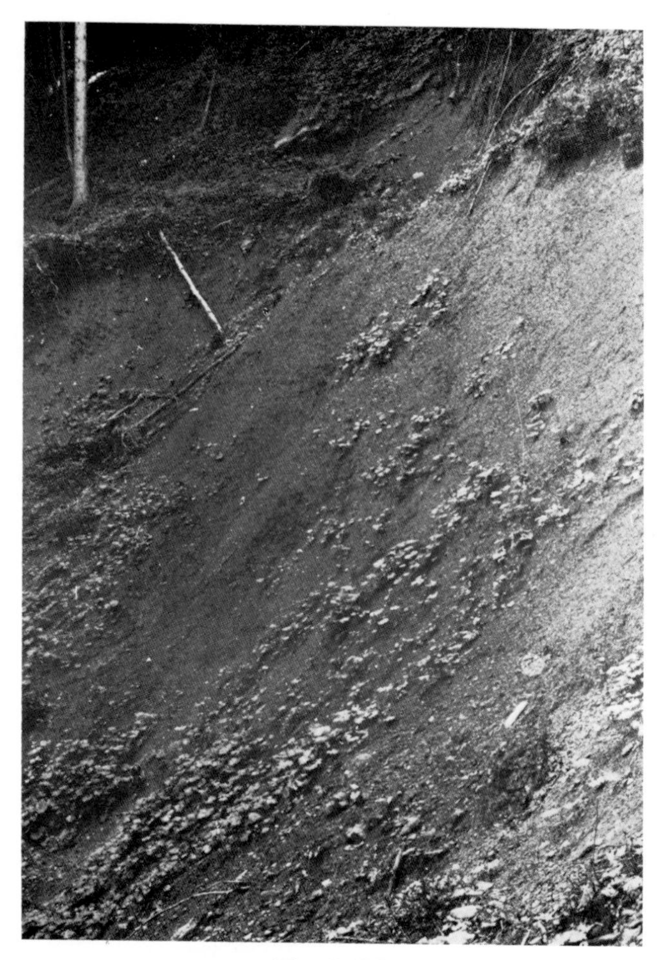

Fig. 5.43B

Fig. 5.43 B. *Tear scar: detail.* A tear scar usually exists over many years. Although the vegetation tries to reestablish itself, it does not seem to be able to take a hold. The biological processes involved are not well known. The picture shows the detail of a tear scar at the head of a clay slide near Schinznach, Canton of Aargau, Switzerland

Fig. 5.43C

Fig. 5.43C. *Tear scar: small slide*. Even small slides leave behind tear-scars as seen here in the vicinity of the railway station on the Ütliberg, Canton of Zürich, Switzerland

Fig. 5.44. *Ground fissures.* Above the region of actual mass movements, small displacements cause ground fissures to arise, as seen above a slide in the Entlebuch Valley, Canton of Lucerne, Switzerland

Fig. 5.45. *Grass slides.* If the slow motion phenomena are confined to the grass layers on a slope, one speaks of grass slides. These are evident in the formation of "terracettes" which become enhanced subsequently by being used by cattle during forage (picture taken on the flank of the Atzmännig, Canton of St. Gall, Switzerland)

Fig. 5.44

Fig. 5.45

Fig. 5.46A

Fig. 5.46A. *Mountain fracture: general aspect.* The slow mass movements on the sides of a crest make it appear that the latter seems "fractured": A trench opens up as seen on the Archenkamm in the Felber Valley, Salzburg Province, Austria

Fig. 5.46B

Fig. 5.46B. *Mountain fracture: lubrication.* Moisture, often in the form of snow, accumulates in the trench at the crest of a mountain; this moisture enters the soil/rock and thereby causes an additional lubrication so that the movements become enhanced. Felber Valley, Salzburg Province, Austria

Fig. 5.46C

Fig. 5.46C. *Mountain fracture: effect on side*. The mass movements on the flanks of a mountain undergoing mountain fracture cause the upper layers of the rocks to break up, as seen in the surficial molasse layers at the Chrüzegg, Canton of St. Gall, Switzerland

Fig. 5.51 A

Fig. 5.51 A. *Rock fall: initiation.* When the stability on a rock wall is somehow exceeded, part of the wall breaks off and falls in a free fall downward, as seen here in Lecco, Como Province, Italy

Fig. 5.51 B. *Rock fall: subsequent events.* After the rock fall reaches the scree apron below, a debris slide is triggered (Lecco, Italy) (after Scheidegger, 1984, 1985 b)

Fig. 5.52 A. *Landslide: major.* As noted above, slope changes may occur quite spontaneously. The most spectacular such instances occur undoubtedly in landslides in which large masses of rock may suddenly break off a mountain ledge and slide downhill with tremendous force. In the slide at Frank, Alberta, a whole village was engulfed in the early morning of April 29, 1903, causing the death of 70 people

Fig. 5.51B

Fig. 5.52A

Fig. 5.52B

Fig. 5.52B. *Landslide: minor*. Even minor landslides can cause damage by blocking roads etc., as seen here on a slide near Choroni, in the coastal range of Venezuela

6 River action

Rivers are very powerful agents in shaping our globe's surface. They act essentially in two fashions: (i) by removing material from its confines, and (ii) by transporting and depositing it. The fundamental processes by which this action is accomplished, are called *river bed processes* (Figs. 6.11 to 6.14).

The removal of material by flowing water from a confining channel, in turn, can again occur in two ways: either the channel is being scoured out and thereby deepened (vertical river action; Figs. 6.21 to 6.23), or the removal occurs on the side. The latter case is referred to as *lateral river action* (Figs. 6.31 to 6.33). A consequence of the lateral river action is that river channels are almost never straight. Contrary to marbles rolling down an inclined plane along the line of steepest descent, water in time scours itself a meandering channel even if the latter should have been straight at the beginning. In soft material the meanders are not stationary; their oscillations leave behind river terraces in this case.

By its erosive action, a river causes a *valley* to be formed. Valleys (Figs. 6.41 to 6.43) show a characteristic structure along their length inasmuch as the headwater region is generally a wide circular area; upon this follows a narrow "gorge section" which is also steep, opening up into a broad flat area of alluvial deposition. This has been taken as the expression of a "catena" principle in geomorphology (Scheidegger, 1986), a catena ("chain") consisting of a relatively flat eluvial section, a narrow, active colluvial section, and a broad, alluvial section at the end. The catena scheme can repeat itself several times over along the whole course of a river. All this is notwithstanding the possibility that the general layout of a series of valleys may have been subject to some tectonic predesign.

Finally, we shall deal in this section with the solution and deposition effects of water (Figs. 6.51 to 6.53) which lead to such features as karst landscapes and sinkholes.

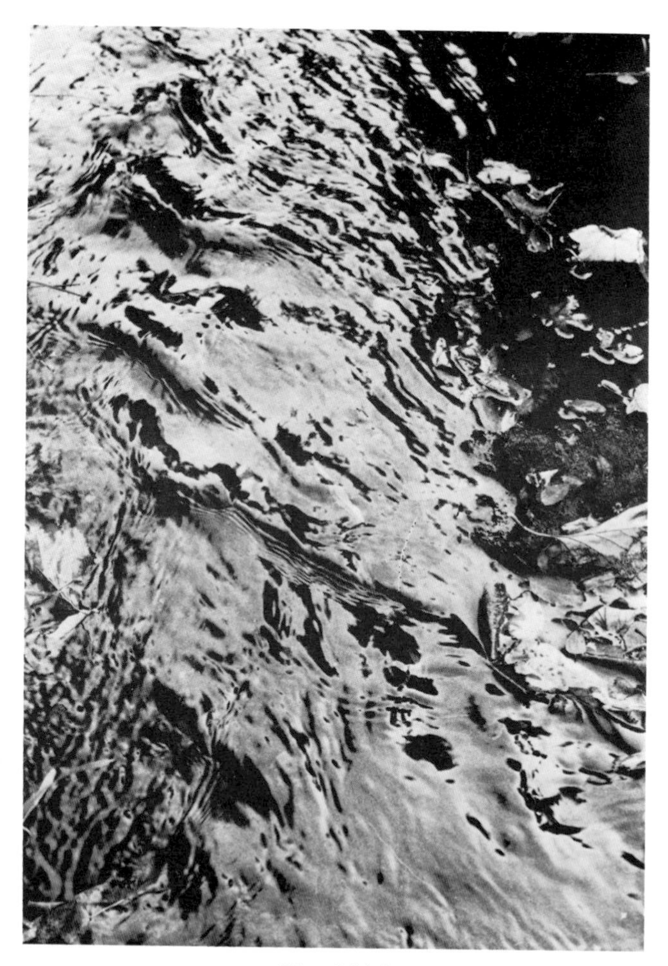

Fig. 6.11A

Fig. 6.11A. *River flow: subcritical flow.* The flow of the water in a river is always *turbulent*. By the term "turbulent" is meant that the local velocity at any one point fluctuates in a random fashion owing to the presence of eddies. The latter can only be described by statistical methods. The turbulence can be seen even in quiet ("subcritical") flow. Note the ripples on the water surface which indicate the presence of turbulence (Halterbach, Vienna Woods, Vienna, Austria)

Fig. 6.11B

Fig. 6.11 B. *River flow: supercritical flow*. The presence of turbulence is immediately evident in "white water". Physicists call such flow "supercritical". It is connected with intense scouring action on the bottom and on the sides of the channel (Gatineau rapids, Québec, Canada)

Fig. 6.12A. *Bed load: submerged.* Whilst the water is flowing in a river, it effects transport of sediments. Parts of these sediments are dragged along at the bottom as *bed load,* parts are carried in suspension and parts are even transported in chemical solution. The bed load may be visible through the water flowing above. Bow river, near Calgary, Alberta, Canada

Fig. 6.12B. *Bed load: lateral.* The bed load becomes visible on the river banks during periods of low water, as seen in the Lech river near Dürnau, Tyrol, Austria

Fig. 6.12A

Fig. 6.12B

Fig. 6.13A

Fig. 6.13A. *Movable bed instabilities: riffles-and-pools.* The instability principle causes a river not to form a uniform channel, but rather a sequence of narrow and wide sections. At the same time, the narrow sections are stretches where the river or brook flows fast (riffles), the wider ones are stretches where a little "pool" appears (Rosenbach in the flysch regions of the Vienna Woods, Vienna, Austria)

Fig. 6.13B

Fig. 6.13B. *Movable bed instabilities: ripples.* The eddies in a river produce a resonance phenomenon with the bed-material. This causes bottom ripples to be formed (Danube river near Dürnstein, Austria, at low water)

Fig. 6.13C. *Movable bed instabilities: bars.* The bed-load in a river forms bars as it is being pushed along. These bars may be on the sides of the river and can be seen when the water level is low. Bow river, near Calgary, Alberta, Canada

Fig. 6.14. *Rocky river bed: scour holes.* On a rocky river bed, the action of the water helped by debris produces scour holes, in consequence of the instability principle: Once the rock surface becomes pitted, pebbles become caught in the hollow, they are swirled around in it, thereby enlarging it. The example shown here was found in a creek whose bed was formed by a lava flow from Mt. Merapi, near Klanggon, Central Java, Indonesia

Fig. 6.13C

Fig. 6.14

Fig. 6.21 A. *Mountain torrents: head section.* The vertical river action manifests itself in the downward erosion in mountain creeks which can evolve into veritable torrents as seen in the Stambach near Bad Goisern, Austria

Fig. 6.21 B. *Mountain torrents: effect in flood plain.* Often, a mountain torrent carries its load of stones far out into the flood plain where it leaves a trail of destruction, as seen here in the vicinity of La Florida in the Colombian Andes

Fig. 6.21A

Fig. 6.21B

Fig. 6.22A. *Gorges and canyons: flat layers.* One of the most impressive examples of vertical river action is seen in the Grand Canyon of Arizona. The Colorado river was present before the uplift of the plateau began, and, as the surrounding territory was slowly rising, it cut itself a tremendous gorge about 2 kilometers deep

Fig. 6.22B. *Gorges and canyons: inner canyon.* In the Grand Canyon of Arizona, there are canyons within canyons. The deepest part in which the river flows is a narrow gorge

Fig. 6.22A

Fig. 6.22B

Fig. 6.22C. *Gorges and canyons: coulee.* Another famous example of vertical river action is seen in the Grand Coulee of Washington State, which was formed by the temporary diversion of the Columbia river during the ice ages. The water course is now empty, and called a "coulee"

Fig. 6.22D. *Gorges and canyons: folded mountains.* The canyons shown above were formed by large rivers in essentially flat-lying strata. In folded mountain country, the vertical river action produces a wild, narrow gorge. The example shown here is the gorge of the Shoshone river at its outflow from Yellowstone National Park, Wyoming, USA

Fig. 6.22C

Fig. 6.22D

Fig. 6.22E

Fig. 6.22E. *Gorges and canyons: narrow breakthrough.* When a river breaks through a rock layer, a narrow canyon is created as shown here in the Glen Helen, N.T., Australia

Fig. 6.23A

Fig. 6.23 A. *Waterfalls: general genesis.* A particular type of vertical river action results in waterfalls. Thus, any resistant rock layer encountered by the river will cause a waterfall, since the erosion below is more rapid than at the top (Ötschergräben, Lower Austria)

Fig. 6.23 B. *Waterfalls: in plains areas.* In flat lying strata a waterfall may be caused by the presence of an erosion scarp across the course of a river. This is the case of Niagara Falls: hard, Middle Silurian dolostones at the level of Lake Erie are underlain by standstones and shales of the cataract group belonging to the lowermost Silurian. The countryside drops along the scarp to the plane consisting of older Ordovician shales surrounding Lake Ontario. The Niagara river carrying the outflow from Lake Erie drops over this scarp, eroding the soft rock below the hard dolostones, causing receding waterfalls which regenerate themselves by the erosion at their foot

Fig. 6.23 C. *Waterfalls: in plains areas.* Niagara Falls offered a rather strange sight when they were artificially "turned off" for "repairs": all the water was routed through the power station whilst the debris at the foot of the fall were removed by bulldozers and cranes—the idea was to restore the whole height of the falls for the tourists. The debris had accumulated because the flow over the falls had been reduced for power generation; thus it was no longer sufficient to carry off the bed load and the falls were in danger of becoming completely clogged and developing into mere rapids

Fig. 6.23B

Fig. 6.23C

Fig. 6.23D

Fig. 6.23D. *Waterfalls: in plains areas.* The shape of the rocks beneath a waterfall cannot be seen while water is flowing over them. However, this can be done in the Dry Falls of Washington. Here the Columbia river formed a waterfall of tremendous size at one time. Eventually the river sought itself a different course, leaving its former falls high and dry. The astonished visitor can now admire them and study in detail the rocks beneath a waterfall

Fig. 6.23E

Fig. 6.23E. *Waterfalls: mountains-tectonic.* In the Alps, great waterfalls look like bridal veils drooping over a precipice. The tremendous height and the relatively small discharge give them this appearance. Shown here are the Giessbach Falls in the Canton of Berne, Switzerland

Fig. 6.23F. *Waterfalls: volcanic*. The Daranak Falls on the Tanay river (Rizal Province, Luzon, Philippines) have been caused by the deposition of a pyroclastic sill in a volcanic eruption. The sill constitutes a hard layer which is resistant to erosion

Fig. 6.23G. *Waterfalls: end stage*. Eventually, as waterfalls recede, they may reach a point above the hard rock layer that was the cause of their formation. Then, they disappear into *rapids* (Bow Falls, Banff, Alberta, Canada)

Fig. 6.23F

Fig. 6.23G

Fig. 6.31 A. *Bank erosion: hard material.* The pictures shown above concern the vertical action of rivers. However, much more important is their lateral (sideways) action. In a bend, a river can attack even solid rock (Bow river, near Cochrane, Alberta, Canada)

Fig. 6.31 B. *Bank erosion: sand.* In soft material, the river bank may consist of sand that gets deposited and washed away, depending of the instantaneous currents: this is the effect of a dynamic equilibrium (arm of the Danube river, near Dürnstein, Austria)

Fig. 6.31A

Fig. 6.31B

Fig. 6.31C

Fig. 6.31C. *Bank erosion: pebbles.* Instead of sand, pebbles may accumulate, depending on the local near-bank currents (Danube river, near Dürnstein, Austria)

Fig. 6.31D

Fig. 6.31D. *Bank erosion: mountain stream.* The lateral erosion at the foot of a steep bank of Bündner schist in mountain country (Schraubach valley, Canton of Grisons, Switzerland) causes a slide and a tear scar to develop higher up on the slope (after Scheidegger, 1986)

Fig. 6.32A. *Meanders: in mountainous areas.* The course of a river or brook is never straight: it always forms a sequence of bends called *meanders.* In mountainous country, the meander course rounds the rock spurs alternating from the two sides. Shown here is the Guanare river near Boconó, Est. Trujillo, Venezuela

Fig. 6.32B. *Meanders: in medium activity areas.* In a medium activity flysch zone, a brook forms miniature meanders, as shown here by the Rosenbach in the Vienna Woods, Vienna, Austria

Fig. 6.32A

Fig. 6.32B

Fig. 6.32C. *Meanders: in plains areas.* On a plain the meanders of a major river are of tremendous size: The loops swing back and forth across the plain, generally never remaining in the same place for very long periods (unless interfered with by man). Often, a meander loop is cut off by the river short-circuiting itself; the cut-off loop is called an *oxbow lake*. The Susitna river near Anchorage, Alaska, USA, shows these features

Fig. 6.32D. *Meanders: incised.* If a flood-plain is underlain by some harder rock, and if the meanders attack this harder rock, then the curves of the river become *incised.* They are then much more permanent than those in an ordinary flood-plain. In such a case, one speaks of *incised meanders*. The rivers in the Painted Desert of Arizona form long strings of incised meanders

Fig. 6.32C

Fig. 6.32D

Fig. 6.32E. *Meanders: formation of a natural bridge.* If incised meanders touch each other, a natural bridge may be formed. The picture shows Ayers Park natural bridge in Wyoming, USA

Fig. 6.33. *River terraces.* The oscillation of meander loops gradually lowers the flood plain of a river. Inasmuch as this process returns to a previously occupied place only after some time, the new level has been lowered by a finite amount and, against the old level, a terrace is formed. The genesis of river terraces is particularly clearly seen in the loess area near Lanzhou, Gansu Province, China

Fig. 6.32E

Fig. 6.33

Fig. 6.41 A. *Valleys, upper reach: general aspect.* The upper reach of a river, i.e. the headwater region, is generally found in mountain country. The catchment area, the "eluvium" in the catena theory, is a widened cirque that can be of impressive dimensions as in the Bighorn mountains of Montana shown here

Fig. 6.41 B. *Valleys, upper reach: catchment cirque.* From the ground, the catchment cirques present the aspect of a big amphitheater (Fallbach catchment area, Malta valley, Carinthia, Austria)

Fig. 6.41A

Fig. 6.41B

Fig. 6.41 C. *Valleys, upper reach: illustration of microcatena.* Even in mountainous areas, small flood plains may be found, illustrating the catena-principle on a quasi-microscale. These flood plains in mountain areas may become fairly wide. Naturally, the debris covering them is coarse and angular in contrast to flood plains in flat country. (Shown is the Elbow river valley, near Bragg Creek, Alberta, Canada)

Fig. 6.42. *Valleys, middle reach.* In the middle reach of a river the gradient is not as steep as in the headwater region. But again, catena-elements (flat-steep-flat) may follow each other. Shown here is the middle reach of the Murrumbidgee river near Canberra, A.C.T., Australia

Fig. 6.41C

Fig. 6.42

Fig. 6.43. *Valleys, lower reach.* As a river enters a plains area, it becomes broad and sluggish. The Mississippi near Ste. Genevieve, Missouri, is a good example

Fig. 6.44. *Tectonic design of valley trend.* rivers often follow the trend of a tectonic feature, such as an escarpment or a rift. In this fashion, the Saguenay river cuts through the Laurentian hills of Québec

Fig. 6.43

Fig. 6.44

Fig. 6.51 A. *Solution effects of water: dissolution.* Water has a solution effect on water-soluble materials. This is particularly well seen in the coral limestone in a quarry near St. Anne on Guadeloupe (F.W.I.)

Fig. 6.51 B. *Solution effects of water: sink holes.* The solution effects of water can proceed to the formation of sink-holes and cave-ins. The hole shown here was formed in gypsum in the Tränenbach gully, Vorarlberg, Austria

Fig. 6.51A

Fig. 6.51B

Fig. 6.51 C. *Solution effects of water: consolidation*. Sink holes can be formed even in materials that are not intrinsecally water-soluble. The sink hole shown here was found in a volcanic tuff near Gimi Junction, Plateau State, Nigeria. It has been caused by washout and consolidation

Fig. 6.52 A. *Karst: general aspect*. The solution effects of water tend to enhance any crack or joint in a carbonate terrain. The resulting landscape is then that of a "karst plateau". The picture was taken in the Hochkönig area, Salzburg Province, Austria

Fig. 6.51C

Fig. 6.52A

Fig. 6.52B. *Karst: fluting*. Karst areas show often the characteristic effect of fluting: Rills become enhanced and further leached out. Shown here are the flutes in limestones at the Spanegg, Canton of Glarus, Switzerland

Fig. 6.52.C. *Karst: sink hole*. The karstification can proceed to the formation of veritable sink holes, as seen here on the plateau of the Rax in Lower Austria

Fig. 6.52B

Fig. 6.52C

Fig. 6.53A

Fig. 6.53A. *Deposition effects of water: tufa terracettes.* Material dissolved in water can also be redeposited under certain conditions: Calcium carbonate is more soluble in cold water than in hot water. Thus, as spring water becomes warmed up while it is trickling down a slope, tufa is deposited. According to the instability principle, this occurs in the form of terracettes. Shown here are the spring-tufa teracettes near Kollbrunn, Canton of Zürich, Switzerland

Fig. 6.53B

Fig. 6.53 B. *Deposition effects of water: stalactites.* Peculiar forms are the stalactites that are formed by the redeposition (in this case) of sodium sulfate in a mine gallery at Bad Ischl, Austria

7 Large bodies of water

It is well known that most of our globe's surface is covered by water. Water is not only found in the oceans, but also in other large bodies such as lakes and inland seas. The morphology of such features shows some interesting traits (Figs. 7.11 to 7.14).

The interaction of water-covered areas with land areas is everywhere approximately the same, whether the former be lakes or oceans: it occurs mainly because of the waves and currents that are present in any large body of water (Figs. 7.21 and 7.22). According to the nature of the waves and currents the shore line may develop bays, beaches or steep cliffs (Figs. 7.31 to 7.34). Particularly interesting phenomena occur at river mouths where estuaries, deltas or fyords may be formed (Figs. 7.41 to 7.43).

Contrary to the dynamics of shore lines which is similar on lakes and oceans, truly deep-water phenomena occur only in oceans. Unfortunately, because of the relative inaccessibility of the oceanic abysses, our knowledge about them is to this day still somewhat scanty.

Fig. 7.11. *The sea surface.* More than two thirds of the earth are covered by the sea. The morphology of the earth's surface is hidden beneath the limitless expanse of the ocean which may only be broken by the eerie appearance of an iceberg in the cold waters of the North Atlantic

Fig. 7.12. *Ocean bottom and marine orogenesis.* The tectonic processes at the ocean bottom are entirely different from those on land: Generally, the relief at the ocean bottom is caused by the presence of "constructive" plate boundaries: the lithosphere is newly created at such boundaries by the upwelling of mantle material; the consequence of this process is the development of midocean ridges. These can be seen in certain places where they reach above the surface of the sea, such as in Iceland

Fig. 7.11

Fig. 7.12

Fig. 7.13A. *Eustatic changes: coasts of emergence.* Relative changes of sea level with regard to the land level are called "eustatic changes". Depending on their direction, such changes give rise to the two fundamental types of coasts: coasts of emergence and coasts of subemergence. Coasts of emergence are fairly rare at the present time. An example which is often quoted (although this is by no means absolutely certain) is the Gulf coast of North America which is characterized by large off-shore sand-bars. The city of Galveston, Texas, is built on such an offshore sand-bar

Fig. 7.13B. *Eustatic changes: coasts of subemergence.* Most coasts present shorelines of subemergence, because the sea level seems to be rising everywhere owing to the melting of land-locked ice masses which is still going on since the last ice age. Thus, on most shorelines, one finds drowned valleys and similar features. An example is the coast of Cape Breton, Nova Scotia, Canada: large inlets form good harbors

Fig. 7.13A

Fig. 7.13B

Fig. 7.14. *Islands*. The sea, dull and grey for most of its extent, may suddenly present views of incredible beauty. How relieved the early explorers must have felt when they saw the sun setting behind a long-sought West-Indian island. (Photograph was taken at sea 16 km off Barbados)

Fig. 7.21 A. *Waves: on open sea*. The waves on the open sea are created by storm centers from which they may travel over very large distances. In a storm, the wave crests form an interference pattern that can be dealt with by the methods of statistical mechanics. The picture was taken in the North Atlantic

Fig. 7.14

Fig. 7.21A

Fig. 7.21 B. *Waves: shoaling*. The most important interaction between large bodies of water and land occurs through the action of shoaling waves. The surf rolling onto a beach presents a picture of great beauty (Jacksonville Beach, Florida, USA)

Fig. 7.21 C. *Waves: breaking*. The tremendous destructive action of breaking waves becomes evident from an inspection of the indicated picture. Every eighth wave crest is hurled with tremendous force against the rock (which is about 10 m high) shown in the foreground. The particular rock has a macabre connotation inasmuch as many jilted lovers have committed suicide by climbing it. Since only every *eighth* wave breaks with the force shown, there is time to ascend the rock (the Black Rock of Louisbourg, Nova Scotia, Canada) between cataclysms

Fig. 7.21B

Fig. 7.21C

Fig. 7.22. *Currents.* In addition to waves, *currents* also have a direct mechanical effect upon the land. It is difficult to photograph ocean currents, but their effect may be seen far out at sea. Shown here are patches of sea weed that had been transported far out into the Atlantic by the Gulf Stream

Fig. 7.31 A. *Steep coast: clay cliffs.* Like a wall on a mountain side, a steep coast is generally the result of decay at the foot of the cliff, particularly if the material is soft, as seen here in the clay bluffs near Scarborough on Lake Ontario (Canada) (after Scheidegger, 1985 b)

Fig. 7.22

Fig. 7.31A

Fig. 7.31B. *Steep coast: rock cliff.* A rock cliff may be the result of some tectonic process: evidently, a sheer drop of a rock wall due to tectonic effects causes a particular type of steep coast as that encountered at the island of Capri (Bay of Naples, Italy)

Fig. 7.31C. *Steep coast: with beach.* Whether a coast is steep or flat depends in essence on the tectonic background, however, in most, instances, the actual appearance of a coast is influenced by erosional processes. In general, surf action tends to increase the declivity so as to produce a cliff; the eroded material accumulates at the bottom of the cliff if the nearshore circulation is such that the formation of a beach is possible. A classic example of this type of erosion is seen in the Falaises of Normandy (France)

Fig. 7.31B

Fig. 7.31C

Fig. 7.31 D

Fig. 7.31 D. *Steep coast: with shore platform.* A cliff recedes by the decay induced by wave action at its foot. If a beach cannot be formed because of the particular near-shore circulation system that is present, a platform is created. The picture was taken near Nasugbu, Batangas province, Luzon, Philippines

Fig. 7.32A

Fig. 7.32A. *Shallow coast: rocky platform.* A shore platform also develops on a shallow coast if the hinterland does not present any considerable relief. This is seen in the St. Lawrence estuary near Ste. Flavie, Québec, Canada

Fig. 7.32B. *Shallow coast: beach genesis.* The form of a shallow coast depends on the nearshore circulation system. The constant supply of sand by longshore drift in dynamic equilibrium with the erosive wave action causes a sandy beach to form (Gulf of Venezuela, near El Moján, Est. Zulia, Venezuela)

Fig. 7.32C. *Shallow coast: Shingle beach.* Not all beaches are covered with sand, some of them consist of much larger material, viz. of stones, possibly several inches in diameter. Such beaches are called "shingle beaches". An example is encountered on the shores of Lesser Slave Lake in Alberta, Canada

Fig. 7.32B

Fig. 7.32C

Fig. 7.32D. *Shallow coast: beach sand ripples.* Resonance effects between the tidal currents and the sand cause sequences of ripples to form in the latter. Beach near El Moján on the Gulf of Venezuela

Fig. 7.32E. *Shallow coast: beach sorting mechanism.* The positioning of fine and coarse materials on a beach is the result of a sorting mechanism in consequence of a dynamic process in which wave action and nearshore circulation is involved; it is to this day not yet completely understood. A peculiar form of sorting of beach material is seen in the above picture where a large number of dead fish have been washed up in a line along a beach (Lake Ontario Beach, near Toronto, Canada)

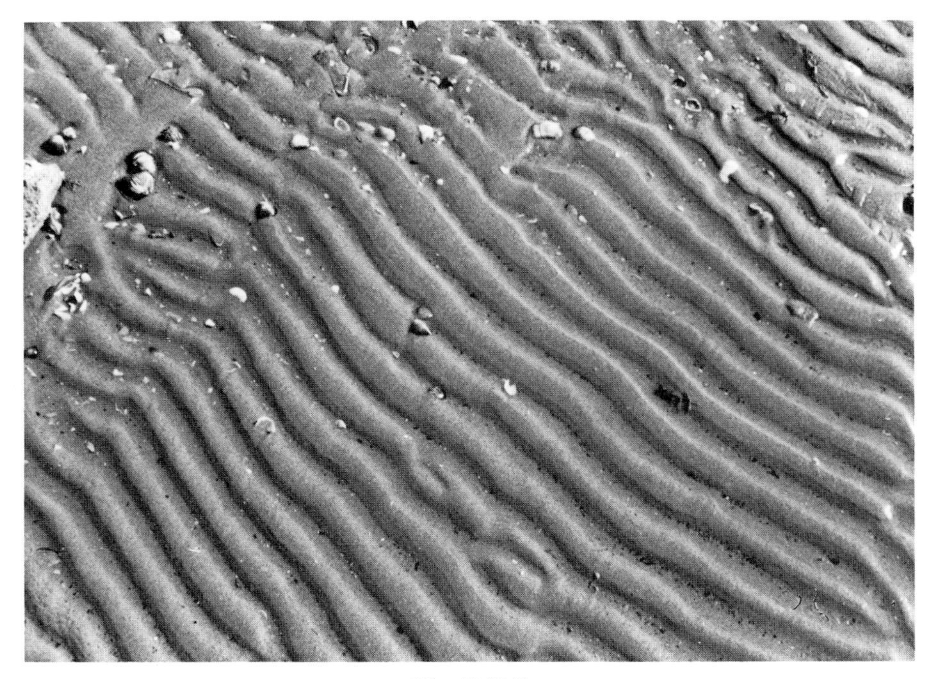

Fig. 7.32D

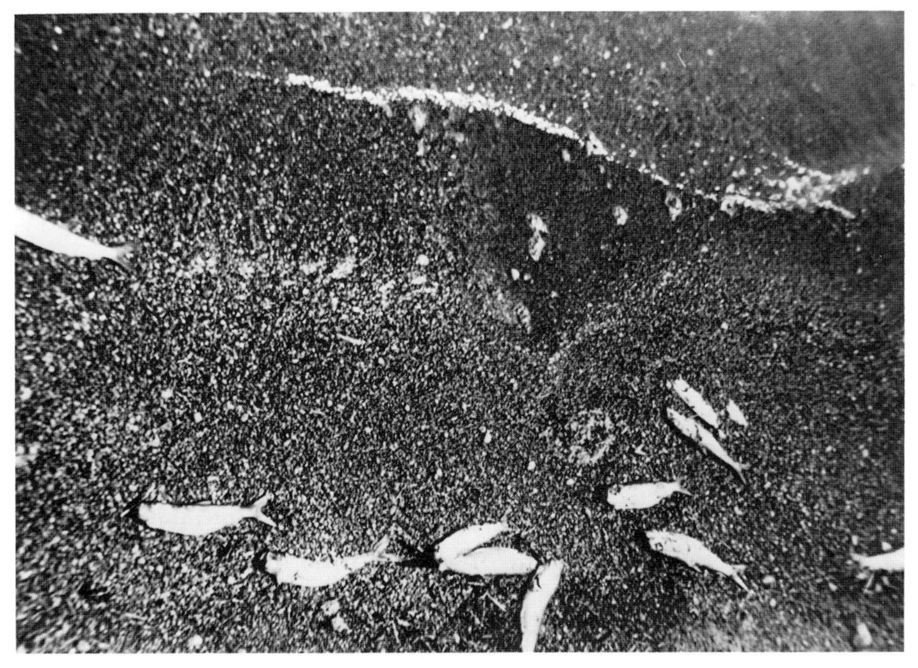

Fig. 7.32E

Fig. 7.32F. *Shallow coast: beach berm.* From the open water side a beach rises to a crest (berm crest) which corresponds to the maximum storm wave swash of recent history. Behind the berm crest, there is generally a backshore depression, often filled with water. The picture shows the berm crest on the south (lake) side of Toronto Island in Lake Ontario, Canada

Fig. 7.32G. *Shallow coast: mud flats.* If the shore material is very fine, mud flats develop in the tidal range, as in the wadden areas on the North Sea coast of Schleswig-Holstein, Federal Republic of Germany

Fig. 7.32F

Fig. 7.32G

Fig. 7.32H. *Shallow coast: mud ripples.* On mud flats, the tidal flow produces ripples similar to those produced by resonance effects on a river bed (mud flats near St. Peter-Ordning, Schleswig-Holstein, Federal Republic of Germany)

Fig. 7.32I. *Shallow coast: mangroves.* In brackish lagoons under a humid tropical climate there arises a typical "mangrove" vegetation consisting of Rhizophora mangle and Avicennia germinans which form the actual shore line. Shown here is the mangrove coast on the Laguna de Sinamaica, Est. Zulia, Venezuela

Fig. 7.32H

Fig. 7.32I

Fig. 7.33 A. *Coral reef: general aspect.* Turning now our attention to truly oceanic phenomena, it is indubitable that coral reefs represent some of the most striking features. Many tropical islands are ringed by such reefs which form a barrier on which the surf breaks. This pattern is very clearly seen from the air, when looking upon the coast of the island of Petite Terre (F.W.I.): the white line of breakers indicating the reef follows the coast line at some distance

Fig. 7.33 B. *Coral reef: lagoon.* Between the reef and the coast, there is a lagoon. Looking straight across the lagoon from the beach, one just barely recognizes the reef edge as a fine line on the horizon (Sanur Beach lagoon, Bali, Indonesia)

Fig. 7.33A

Fig. 7.33B

Fig. 7.33C. *Coral reef: platform.* The coral reef forms a platform which is seen at low tide. (Matabungkay, Batangas Province, Luzon, Philippines)

Fig. 7.33D. *Coral reef: coral.* The coral reef is built up by coral of which pieces may break off and can be found on the beach. (Sanur Beach, Bali, Indonesia)

Fig. 7.33C

Fig. 7.33D

Fig. 7.34A. *Shore erosion: hard material.* Coastal erosion may produce some very striking features. Shown here is a bay scooped out by the sea from the island of Corfu, Greece

Fig. 7.34B. *Shore erosion: soft material.* The previous picture showed coastal erosion on solid rock. In soft material, somewhat different features result which remind one very much of the effects caused by the sideways erosion of rivers: the coast recedes partly due to the direct attack by the sea, partly due to wastage. The Scarborough Bluffs (near Toronto) on the Lake Ontario shore remind one of badlands

Fig. 7.34A

Fig. 7.34B

Fig. 7.34C. *Shore erosion: bay.* On shallow coasts consisting of soft materials (sand), local currents scoop out shallow bays. On some bays, the currents tend to close off the entrance from the sea by the deposition of sand in the form of a spit. An inflow of fresh water into the bay, however, will keep a gap open at high tide. Shown here is the spit on Deewhy Bay, near Sydney, Australia

Fig. 7.34D. *Shore erosion: Structural background.* Many bays are not caused by coastal erosion, but are caused by structural features. Goose Bay in Labrador is a "bay" of this type

Fig. 7.34C

Fig. 7.34D

Fig. 7.41. *Estuary*. A type of "bay" may be caused by a river emptying into the sea. If longshore currents are present, the material carried by the river into the sea is removed and the river creates for itself an estuary. Shown here is the estuary of the Moose river near Moosonee, Ontario, Canada

Fig. 7.42. *Delta*. If no longshore currents are present, a river dumps all its bed load near its mouth and thereby builds up a delta. (Delta of the Thiamis river on the Adriatic coast of Greece)

Fig. 7.41

Fig. 7.42

Fig. 7.43

Fig. 7.43. *Fyord.* The particular form of a river mouth may not only depend on the nearshore circulation, but may also be influenced by the structure of the area. A fyord is always of the form of an estuary although at its head, a lot of material may be deposited by a river. Shown her is a fyord on the west coast of British Columbia, Canada

8 Niveal features

Niveal features are due to the action of snow and ice. Snow, if it accumulates, consolidates into ice which may form giant streams proceeding downhill in a manner similar to that of rivers. These ice streams are called glaciers. Figs. 8.11 to 8.13 deal with the phenomenology of snow and ice fields and glaciers.

Glaciers have a pronounced effect upon the earth's surface: they erode their bed, they sculpture mountain sides and they deposit material in certain areas. The erosional acitivity of glaciers manifests itself in the scouring out of valleys and mountain sides in a characteristic fashion, the depositional activity in the dumping of various types of features (Figs. 8.21 to 8.23).

The extent of the land areas of the world that are glaciated to-day is relatively small: a large continental ice mass exists presently in Antarctica, another in Greenland. Otherwise, glaciers are found only in high mountain country. However, at various periods during the earth's history, large parts of the world were extensively glaciated. These periods are called "ice ages"; the latest of them occurred during the Pleistocene (ending about 13 000 years ago). In it, North America, Central and Northern Europe, Northern Siberia, the Alps, the Himalayas, the Andes, Patagonia, New Zealand, Tasmania as well as all the presently glaciated areas, were covered by large masses of ice. Another ice age occurred in the Upper Paleozoic epoch (about 200 million years ago), others still earlier. It is neither clear what causes ice ages, nor is it clear whether the latest (Pleistocene) ice age has truly been terminated or whether, at present, we are simply in an interglacial stage of the latter; such interglacial stages are a common feature of ice ages in general. The most recent ice age has had a very pronounced effect upon the morphology of many areas, inasmuch as the results of glacier action are still plainly visible in many now unglaciated areas ("periglacial features", Figs. 8.31 to 8.34).

Fig. 8.11 A. *Snow fields: ordinary snow surface.* As the snow falls, it obtains generally a rippled surface because of the wind action and because of local recrystallization: the ripples (with a wave length of a few cm) correspond to the instability principle of geomorphology. Scene from the Vienna Woods, Austria, in winter

Fig. 8.11 B. *Snow fields: drifts.* On open, flat areas, the snow ripples become drifts. These correspond again to the instability principle, but the wave lengths are now of the order of 10 meters or so. Scene from the Vienna Woods in winter

Fig. 8.11A

Fig. 8.11B

Fig. 8.12A

Fig. 8.12A. *Glacier morphology: origination.* As a snow field grows and as the snow recrystallizes into ice, it starts to push forward into existing valleys, thereby forming a "glacier". Thus, glaciers usually have the form of "fingers" protruding from a large snow field. Shown here is the Worthington glacier, near Valdez, Alaska, USA

Fig. 8.12B

Fig. 8.12B. *Glacier morphology: valley glacier*. Glaciers behave much like rivers flowing in a valley. Thus, one can distinguish elements of a "catena": flat (eluvial)-steep (colluvial)-flat(alluvial) sections; the last flat section may act as the beginning of the next catena. The example shown is from the Steinlimi glacier in the Canton of Berne, Switzerland

Fig. 8.12C. *Glacier morphology: ice falls.* In a steep section, a glacier may develop "ice falls" corresponding to waterfalls in rivers. Shown here are the ice falls in the Steinlimi glacier in the Canton of Berne, Switzerland

Fig. 8.12D. *Glacier morphology: tongue.* A glacier generally ends in a tongue with a lake in front. The tongue is concave when the glacier is retreating, convex when it is adavancing as is the case in the Stein glacier shown here (Canton of Berne, Switzerland)

Fig. 8.12C

Fig. 8.12D

Fig. 8.12E

Fig. 8.12E. *Glacier morphology: glacier door*. The water at the end of a glacier comes out of an ice cave, usually called "glacier door". (Athabasca glacier, Alberta, Canada)

Fig. 8.12F

Fig. 8.12F. *Glacier morphology: hanging glacier*. Instead of flowing in a valley, a glacier may issue from an ice field and proceed downward without reaching the bottom of the main valley. In this case, it is called a "hanging glacier". As an example, we show the Andromeda glacier in the Columbia icefields in Alberta, Canada

Fig. 8.13A

Fig. 8.13 A. *Ice cap: continental.* Not all glaciers rise in high mountain country. Shown here is the western edge of the Greenland ice cap, which is of continental dimensions. Much of the northern hemisphere was covered by ice caps of this type during the last ice age. Not many years ago, great heroism was required to visit and traverse the Greenland ice cap. Today, daily commercial jet flights cross the center of Greenland in a few hours

Fig. 8.13B

Fig. 8.13B. *Ice cap: mountains*. Another ice cap seen on a commercial Europe-to-California flight is that of Baffin Land (Northwest Territories, Canada). Although not as large as the Greenland ice cap, it is far more impressive because it is centered around a high mountain range

Fig. 8.21 A. *Glacial erosion: U-valley.* Glacier scouring has a pronounced effect upon the shape of valleys: such valleys are eroded so as to attain a U-shaped profile. This is particularly clearly seen in valleys that had been glaciated during the last (Pleistocene) ice age. Shown here is a valley in U-shape in which a lake has been formed: Lake Como as seen from Brunate (Italy)

Fig. 8.21 B. *Glacial erosion: glacial cirque.* On steep mountain sides, the eroding action of ice takes on a peculiar form: The ice mass hanging down over a slope forms a glacial cirque which, after the ice melts, becomes filled with water (kar lake). Shown here is the kar lake on the flank of the Tierbergli, Stein glacier region, Canton of Berne, Switzerland

Fig. 8.21A

Fig. 8.21B

Fig. 8.21C

Fig. 8.21C. *Glacial erosion: formation of a peak*. If several glacial cirques were present at one time on several sides of a mountain, a steep peak will be the final result. (Canadian Rockies, Alberta, Canada)

Fig. 8.22A

Fig. 8.22A. *Moraine: terminal.* Glaciers *deposit* as well as erode material. The best-known glacial deposits are *moraines*: on its tongue, a glacier accumulates and carries along a great mass of debris. After the glacier melts away, this material is simply left in place and forms a barrier across the valley. Shown here is a moraine left by an ice-age glacier on the island of Samsø in Denmark

Fig. 8.22B

Fig. 8.22B. *Moraine: lateral.* Moraines are not only formed at the end of a glacier, but also on its sides, as is evidenced here by a lateral moraine left by a glacier near Zermatt, Canton of Valais, Switzerland

Fig. 8.23

Fig. 8.23. *Erratic blocks*. On occasion, glaciers transport large blocks for considerable distances. This happened particularly during the last ice age. After the melting of the glaciers, the blocks are left as erratic blocks, like the "Erdmannlistein" found near Bremgarten, Canton of Aargau, Switzerland

Fig. 8.31. *Patterned ground: frost polygons.* Turning now to periglacial features, we note that the freezing and thawing action in a niveal area causes the ground to break into polygonal patterns, on which the detritus tends to collect. Shown here are frost polygons near the Col de Morcle, Canton of Valais, Switzerland (2800 m)

Fig. 8.32A. *Periglacial deposits: drumlins.* Drumlins are hills of glacial drift which approximate the form of an elongated ovoid. They seem to be streamlined bodies sculptured by ice moving across them; after the ice has melted, they remain part of the periglacial landscape. The picture shows a drumlin near Kärselen, Canton of Berne, Switzerland

Fig. 8.31

Fig. 8.32A

Fig. 8.32B. *Periglacial deposits: kames.* In contrast to drumlins, kames have been deposited not under the ice but against the ice margin. Therefore, they are generally asymmetrical, like the kame near Baden, Ontario, Canada, shown here

Fig. 8.32C. *Periglacial deposits: prairie mounds.* Prairie mounds are round hills with a depression (lake!) in the center; the circular material is glacial till. They are presumed to have been formed by the presence of a large block of ice in the center against which the till was piled up. After the melting of the central ice block, the depression was formed. The picture was taken near Wayne, Alberta, Canada

Fig. 8.32B

Fig. 8.32C

Fig. 8.33 A. *Glacial humps: roche moutonnée.* The ice flowing over its bed forms regions of instability. This may express itself in the formation of a rocky hump, called "roche moutonnée" (Rundhöcker). An example is shown here from the Hochkönig region, Salzburg province, Austria

Fig. 8.33 B. *Glacial humps: pingo.* Another type of "glacial hump" is the "pingo": ice lenses in the ground that cause a conical mound to be formed. Pingoes occur mainly in the Arctic where they may reach a height of 100 m. They can also be found at high altitudes in middle latitudes as demonstrated by a small pingo found at the Ferdenpass (2000 m) near Leukerbad, Canton of Valais, Switzerland

Fig. 8.33A

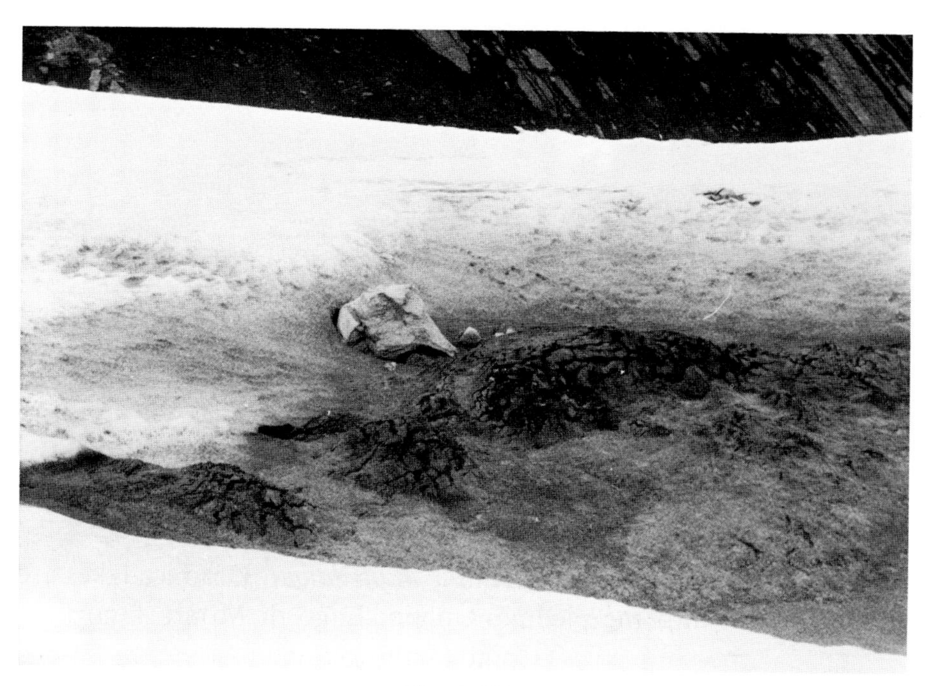

Fig. 8.33B

Fig. 8.34A. *Periglacial dead-ice: lake.* Lakes are formed not only by glaciers dumping moraines which dam the water up, but also by relinquishing masses of "dead ice" in glacially-scoured hollows. On a small scale, this is seen here in the lake formed in front of two roches moutonnées on the Seebenalp in the Flumserberge, Canton of St. Gall, Switzerland

Fig. 8.34B. *Periglacial dead-ice: lake in piedmont.* Dead ice lakes are found far out into the piedmont plain: Lago di Montórfano near Como, Italy

Fig. 8.34A

Fig. 8.34B

9 Desert features and related phenomena

Desert features are characterized by the fact that they arise in dry climates. Under such conditions, it is the absence of water (and ice) that is the most important fact. Thus, wind remains the essential exogenic agent. The action of wind consists in the transportation of loose materials, mostly sand grains, sometimes of dust particles, and in the abrasion of surfaces by blown sand.

In fact, wind can be a powerful geomorphological agent under certain conditions. Although this is mainly manifest in deserts, there are also other instances in which wind action can be important, mainly in the genesis of some coastal dunes and in loess formation. It is convenient to treat such forms in conjunction with the "aeolian effects" (Figs. 9.11 to 9.13) occurring in deserts.

Inasmuch as the climate in deserts is generally not only arid, but also hot, the evaporation of water is very high. This causes the formation of a peculiar "evaporite" environment in which solutes (such as gypsum or salt) contained in water are precipitated out and thus play a significant role in the shaping of the earth's surface (Figs. 9.21 to 9.23).

Finally, rocks exposed to the heat of the day and to the abrasive action of the sand are degraded in a peculiar way which may lead to inselbergs and other specific desert features (Figs. 9.31 and 9.32).

Fig. 9.11. *Clouds*. The blowing wind, like flowing water, is generally in a state of turbulence. Just like the ripples on the surface of a river, "ripples" may exist in a wind stream. Such ripples are particularly noticeable if one is flying over them: suddenly, the upper surface of the clouds forms waves with a constant wave length over great distances (clouds above Venezuelan coast)

Fig. 9.12A. *Dunes: field in desert*. In an aeolian landscape, the presence of sand is a most characteristic feature. Sand, of course, is the ultimate result of the contrition of the rocks of the area. As the sand is blown over the desert by the wind, it tends to accumulate in wandering mounds which are called *dunes*. The dunes may cover areas of thousands of square km in extent as in the Great Erg shown here (Algerian Sahara)

Fig. 9.11

Fig. 9.12A

Fig. 9.12B. *Dune: barchan dune in desert.* The dunes are generally crescentic (i.e. they have the form of a crescent or sickle) because the edges advance faster than the center during the wandering process. On the lee side of each dune, there is a conical slip face. This type of dune is called a "barchan dune" (Algerian Sahara)

Fig. 9.12C. *Dune: compacted.* When dune sand becomes compacted, it acquires the aspect of a sandstone. This has happened with aeolian volcanic sand in the extremely dry climate of Lanzarote, one of the Canary islands (after Scheidegger, 1978)

Fig. 9.12B

Fig. 9.12C

Fig. 9.13

Fig. 9.13. *Aeolian ripples*. The desert wind does not only cause large-scale features such as dunes, but also small-scale ripples in the sand. These ripples are present everywhere and have a wave length of about 20 cm (Algerian Sahara)

Fig. 9.21 A

Fig. 9.21 A. *Evaporite formation: gypsum crystals.* Thy dryness in a desert allows strange types of crystals to grow. Shown here is a specimen of "desert roses" which was found in the Sahara. This is a typical evaporite feature

Fig. 9.21 B. *Evaporite formation: in loess desert.* Evaporites are precipitated if the evaporation is faster than the runoff and the water does not reach an outflow from an area. Such conditions prevail in the loess desert near Lanzhou (Gansu Province) in China

Fig. 9.21 C. *Evaporite formation: in volcanic environment.* The crater lake of Taal volcano (Batangas Province, Luzon, Philippines) contains a large amount of sodium nitrate which forms a white deposit on its shore

Fig. 9.21B

Fig. 9.21C

Fig. 9.22

Fig. 9.22. *Evaporite flats.* Under conditions of evaporite formation the water does not even reach the river valley before it evaporates: Thus, evaporite flats are created as seen here near Lanzhou, Gansu Province, China

Fig. 9.31

Fig. 9.31. *Rocky desert, general aspect.* Not all so-called "deserts" owe their morphology to wind action. The Arizona desert (USA) is still mostly a landscape sculptured by flowing water (during cloudbursts). The vegetation (mostly Saguaros and other cacti) is lush compared with that of the Sahara, although it is meager indeed in comparison with that of the arable regions of the world

Fig. 9.32A

Fig. 9.32A. *Desert rocks: inselberg.* In central Australia, in a dry climate, there are some interesting inselbergs, of which Ayers' Rock is the most famous. It consists of an arkose facies of a conglomerate which was formed from Proterozoic and Paleozoic sediments folded during a Devonian orogeny. It is a true monolith with many traces of weathering and leaching by water, which is precipitated at infrequent intervals

Fig. 9.32B

Fig. 9.32B. *Desert rocks: tor.* Oddly shaped inselbergs are called "tor", like the "man and woman" of Kufena Hill, near Zaria, Kano State, Nigeria shown here

10 Volcanic landscapes

In the final chapter of this book, we shall discuss some geomorphological effects of volcanic activity.

The source of volcanic activity is connected with processes occurring in the interior of the earth. The earth, as a whole, appears to be a heat engine in which thermal energy created or contained in the interior is transferred into space; thereby it is doing work.

In accordance with the fact that mountain building and thermal effects are somehow connected, volcanic activity is, on the whole, concentrated in those areas of the world where recent tectonic activity has been going on. This concerns the recent orogenetic belts as well as the midocean ridges.

The most direct manifestation of the thermal activity in the interior of the earth is presented by the volcanoes themselves. Our first concern is therefore with the morphology of such features (Figs. 10.11 to 10.13).

The volcanic action, through the emission of lava, bombs and ash, causes characteristic features in a landscape which can be recognized as "volcanic" ones (Figs. 10.21 to 10.23).

Finally, volcanoes are not the only manifestations of thermal processes inside the earth; other such manifestations are geysers, hot springs and similar phenomena, with their attendant geomorphological effects (Figs. 10.31 to 10.32).

Fig. 10.11A. *Volcano shape: classical.* The classic prototype of a volcano known since ancient times is probably Mt. Vesuvius near Naples in Italy. At the top there is a crater from which ash and molten rock erupt from time to time. Such an eruption destroyed the cities of Herculaneum and Pompeii in historic times

Fig. 10.11B. *Volcano shape: chain.* The classic conical shapes of volcanoes are exhibited by a chain of volcanoes in central Java, Indonesia. The closest one of these is Goreng Sundoro

Fig. 10.11A

Fig. 10.11B

Fig. 10.11 C. *Valcano shape: single cone.* Another famous volcano is the Popocatepetl in Central Mexico. The conical form, of course, is caused by the dumping of material from eruptions which is piled on from the top. The slight concavity of the slopes is caused by rain water erosion in conformity with the selection principle

Fig. 10.12 A. *Volcanic crater: dry.* The crater of a volcano is usually simply a great big hole which has been blown out clean by the last eruption. The crater on Mt. Tabaro in the Taal volcano complex (Batangas Province, Luzon, Philippines) is of this type

Fig. 10.11C

Fig. 10.12A

Fig. 10.12B. *Volcanic crater: filled with lake.* The crater at the top of a volcano is often filled with water to form a crater lake, as seen here at the summit of the Nevado de Toluca in Central Mexico (after Scheidegger, 1985 b)

Fig. 10.12C. *Volcanic crater: collapsed.* The collapse of an eruption center on Taal Volcano (Batangas province, Luzon, Philippines) left a desolate lava field in its place

Fig. 10.12B

Fig. 10.12C

Fig. 10.13A

Fig. 10.13A. *Volcanic plug: basaltic.* Volcanic plugs can arise by the solidification and subsequent exposure of the material in a volcanic vent. If the intrusive material in the former crater vent is basalt, it usually splits into polygonal columns, as seen on the exposed vent of an old volcano: Devils Tower, Wyoming, USA

Fig. 10.13B

Fig. 10.13B. *Volcanic plug: trachytic.* South of Lantang, in Plateau State, Nigeria, is a famous trachytic plug, standing on an eroded plane surface of Cretaceous sediments (Wase Rock)

Fig. 10.13C. *Volcanic plug: dacite*. A dacitic plug (piton) is found in front of the coast of Martinique, F.W.I.: Le rocher du diamant (after Scheidegger, 1982)

Fig. 10.21A. *Lava flow: basalt*. Lava, particularly basaltic lava, moves from the source like a river, as evidenced by the basalt flow from the Pico de Teide, on Tenerife, Canarias

Fig. 10.13C

Fig. 10.21A

Fig. 10.21B

Fig. 10.21 B. *Lava flow: basalt flow-tongue.* The lava flow comes to a stop when the cooling proceeds to a point at which the liquidity becomes lost. If the material is basalt, the columnar cooling joints form a rosette or striking appearance (location is on Tenerife in the Canary islands) (after Scheidegger, 1978)

Fig. 10.21 C

Fig. 10.21 C. *Lava flow: basalt sill.* If the lava is basaltic, it is fairly liquid and forms large sills. On the side of such a sill, the cooling of the basalt produces characteristic columns which have a polygonal cross section. Shown here is the basalt sill at Kegon Falls near Nikko, in Central Honshu (Japan)

Fig. 10.21 D. *Lava flow: basalt field.* Basaltic lava flows occur mainly in the rift areas of the world. They form weird rugged fields into which rivers cut their courses as is the case in the above picture taken near Portuguese Bridge, in the rift zone of Ethiopia

Fig. 10.22. *Volcanic bombs.* At times, a volcano ejects whole blobs of lava which solidify during the flight in a more or less spherical shape. These fall as "volcanic bombs", like the one shown here which was found near Kenang, Plateau State, Nigeria. It came from an eruption of Kereng Hill that took place about 0.5 million years ago

Fig. 10.21D

Fig. 10.22

Fig. 10.23A

Fig. 10.23A. *Lapilli: general.* When the ejected lava "blobs" are rather small, one does not speak of "bombs", but of "lapilli". These correspond to solidified lava droplets of some millimeters in diameter (Gran Canaria) (after Scheidegger, 1978)

Fig. 10.23B

Fig. 10.23B. *Lapilli: stratified.* Very small lapilli can form stratified air deposits as seen here on Mt. Tabaro, Batangas Province (Taal region), Luzon, Philippines. The deposit stems from an eruption that occurred in 1976

Fig. 10.24

Fig. 10.24. *Lahar*. When a crater lake suddenly empties out, it causes a debris flow of water and volcanic material. This is called a "lahar". Shown here is a lahar stream in Basse Terre, Guadeloupe, F.W.I. (after Scheidegger, 1982) (cf. also Fig. 4.13A)

Fig. 10.25

Fig. 10.25. *Caldera.* A volcanic eruption depletes the magma chamber below of material; hence a mass defect arises and the material above collapses in a circular area to form a depression; this is called a caldera. The picture shows the main caldera on the Dieng Plateau in Central Java, Indonesia

Fig. 10.31A

Fig. 10.31A. *Geyser: phenomenology.* As noted, volcanic effects include hydrothermal phenomena. The most spectacular ones of these are geysers which are springs that are quiet for certain periods of time and then spurt out large quantities of water and steam, often with great regularity. Shown here is Old Faithful Geyser in Yellowstone National Park, Wyoming, USA

Fig. 10.31 B

Fig. 10.31 B. *Geyser: geyserite*. The water coming out from the mouth of a geyser cools down very rapidly and loses therefore many of its solute materials. This material precipitates as "geyserite" around the mouth of the geyser building up a mound of such geyserite. The picture was taken in the geyser field on the Dieng Plateau in Central Java, Indonesia

Fig. 10.32

Fig. 10.32. *Hot spring*. Even if a hot spring is not as spectacular as the geysers shown in the previous pictures, it may give rise to remarkable features. Shown here is a "paint pot" in Yellowstone National Park, Wyoming, USA

References

Davis, W. N. (1924): Die erklärende Beschreibung der Landformen, 2. Aufl. Leipzig: Teubner.

Gerber, E. (1969): Bildung von Gratgipfeln und Felswänden in den Alpen. Z. Geomorph. [Suppl.] **8**: 94—118.

Gerber, E., Scheidegger, A. E. (1973): Erosional and stress-induced features on steep slopes. Z. Geomorph. [Suppl.] **18**: 38—49.

Gerber, E., Scheidegger, A. E. (1974): On the dynamics of scree slopes. Rock Mech. **6**: 25—31.

Kohlbeck, F., Scheidegger, A. E. (1977): On the theory of the evaluation of joint orientation measurements. Rock Mech. **9**: 9—25.

Kohlbeck, F., Riehl-Herwirsch, G., Roch, K. Scheidegger, A. E. (1980): In situ Spannungsmessungen an der Periadriatischen Naht in der Ebriachklamm bei Eisenkappel (Kärnten, Österreich). Mitt. Ges. Geol. Bergbaustud. Österr. **26**: 139—153

Penck, W. (1929): Die geomorphologische Analyse. Stuttgart: J. Engelhorn's Nachf.

Scheidegger, A. E. (1958): A physical theory of the formation of hoodoos. Geofis. Pura Appl. **41**: 101—106.

Scheidegger, A. E. (1970): Theoretical geomorphology, 2nd edn. Berlin—Heidelberg—New York: Springer.

Scheidegger, A. E. (1976): Correlation between joint orientation and geophysical stresses in a test area on the Canadian Shield. Rock Mech. **8**: 23—33.

Scheidegger, A. E. (1978): The tectonic significance of joints in the Canary Islands. Rock Mech. **11**: 69—85.

Scheidegger, A. E. (1979): The principle of antagonism in the earth's evolution. Tectonophysics **55**: T 7—T 10.

Scheidegger, A. E. (1982): Vulkanismus. Physik in unserer Zeit **13**, 5: 131—137.

Scheidegger, A. E. (1983): Instability principle in geomorphic equilibrium. Z. Geomorph. **27**, 1: 1—19.

Scheidegger, A. E. (1984): Physik der Naturkatastrophen. Physik in unserer Zeit **15**, 4: 100—109.

Scheidegger, A. E. (1985 a): Garlands of heads. Z. Geomorph. **29**, 2: 223—234.

Scheidegger, A. E. (1985 b): Physik der Naturkatastrophen. Arcus (Inst. f. internationale Architektur-Dokumentation) **1985**, 1: 7—16.

Scheidegger, A. E. (1986): The catena principle in geomorphology. Z. Geomorph. **30**, 3: 257—273.

Scheidegger, A. E. (1987): The fundamental principles of landscape evolution. In: Proceedings, Workshop on Theoretical Geomorphological Models, Aachen, April 2—5, 1986. Catena [Suppl.] (in press).

Scheidegger, A. E. (1988): The dynamics of geomorphic systems. In: Proceedings, Symposium on Dynamic Systems Approach to Natural Hazards, IASH, Vancouver, August 21, 1987. Z. Geomorph. [Suppl.] (in press).

Scheidegger, A. E., Ai, N. S. (1986): Tectonic processes and geomorphological design. Tectonophysics **126**: 285—300.

Scheidegger, A. E., Padale, J. G. (1982): A geodynamic study of peninsular India. Rock Mech. **15**: 209—241.

Strahler, A. N. (1957): Quantitative analysis of watershed geomorphology. Trans. Amer. Geophys. Un. **38**: 913—920.

Terjung, W. H. (1982): Process-response systems in physical geography. Bonn: F. Dümmler.

Thom, R (1972): Stabilité structurelle et morphogenèse. Reading, Mass.: Benjamin.

Index

Abrasion 239
Acadian orogeny 90, 91
Activity, landscape 2, 3, 11, 12–25
Addis Ababa 62, 63
Aeolian features 76, 77, 239–251
Ai 5, 276
Ajanta 104, 105
Alaska 160, 161, 216
Alberta 22, 23, 46, 47, 78, 89, 99, 128, 129, 134, 135, 138, 139, 152–155, 166, 167, 194, 195, 220, 221, 226, 232, 233
Algeria 24, 25, 76, 77, 240, 241–244
Alluvium 22, 23, 131, 217
Alps 12–15, 151, 213
Anatexis 92, 93
Andes 12, 13, 140, 141, 213
Andesite 58, 59
Andromeda glacier 221
Angle of repose 106
Antagonism principle 2, 3, 5, 11 ff.
Antarctica 213
Anticline 28, 29
Appalachians 18, 30, 31, 68, 69
Aqueous erosion 101–105
Archenkamm 124, 125
Arctic 19, 22, 23, 234
Arid climate 1, 11, 16, 17, 20, 21, 24, 25, 239–251
Arizona 16, 17, 142, 143, 160, 161, 249
Arkose 250
Ash 253, 254
Athabasca glacier 220
Atlantic ocean 180, 181, 184, 185, 188, 189
Atmospheric processes 4
Atzmännig 80, 81, 122, 123
Australia 55, 68, 69, 103, 146, 166, 167, 208, 209, 250
Australian Commonwealth Terr. 166, 167
Austria 16, 17, 28, 29, 34, 35, 62, 63, 74, 75, 82, 83, 86–88, 108, 109, 111, 114, 115, 117, 119, 124, 132, 134, 136, 137, 140, 141, 147, 154–156, 158, 159, 164, 165, 170–175, 177, 214, 215, 234, 235
Ayers Park 162, 163
Ayers Rock 24, 25, 250

Babo Shan fault 32, 33
Backshore depression 198, 199
Baden 232, 233
Bad Goisern 114, 115, 140, 141
Bad Ischl 83, 177
Badlands 89, 99, 206, 207
Baffin Land 19, 223
Bali 14, 15, 202, 203
Ball-and-pillow structure 88
Bancroft 92, 93
Bank erosion 154–157
Bar 138, 139, 182, 183
Barbados 184, 185
Barchan dune 242, 243
Barren lands 22, 23
Barrier 202, 203
Basalt 41, 58–60, 61, 66, 67, 104, 105, 260, 262–267
Base level, erosion 1, 22, 23
Basement rocks 66, 67
Basic landscapes 11–25
Basse Terre 65, 270
Bay 206–210
Beach 74, 75, 186, 187, 190–192, 194–199
Bed instability 136–139
Bed load 76, 77, 134, 135, 138, 139, 210
Beijing 32, 33
Belemnite 36
Berm 198, 199
Bighorn mountains 164, 165
Biological effect 78, 120, 122, 123

Black Rock 186, 187
Block, erratic 229
Block faulting 68
Block, on slope 108, 109
Blown sand 239–241
Blue Mountains 30, 31
Bluff, *see* Cliff
Bohemian massif 16, 17, 34, 35, 74
Bomb 253, 266–268
Boudinage 34, 35
Bow River 134, 135, 138, 139, 152–155
Breaking wave 186, 187, 202, 203
Bridge, natural 162, 163
British Columbia 212
Bündner schist 84–87, 157
Butte 20, 21

Cactus 249
Calcite 44
Caldera 271
Canada 19, 22, 23, 28–33, 38–40, 46, 47, 50, 51, 78, 84, 85, 89–93, 99, 128, 129, 133, 138, 139, 148, 149, 152–155, 166– 169, 182, 183, 186–189, 193–199, 206–212, 220, 221, 223, 226, 232, 233
Canary Islands 242, 243, 262–264, 268
Canyon 51, 142–146
Cap, ice 222, 223
Cape Breton 182, 183
Capri 190, 191
Caracas valley 114, 115
Carbonate 172, 176
Caribbean, *see* West Indies
Catastrophe theory 4
Catchment area 104, 105, 164, 165
Catena principle 5, 131, 166, 217
Cave 220
Chemical reduction 70
Chevron fold 32, 33
China 32, 33, 162, 163, 246–248
Choroni 130
Chrüzegg 126
Cirque 104, 105, 164, 165, 224–226
Clay 79–82, 116, 188, 189
Cliff 71, 74, 75, 188–192, 206, 207
Climate, effect 11
Clingmans dome 18
Cloud 240, 241
Cloudburst 249
Coast 74, 75, 179, 182, 183, 188–201
Col de Morcle 230, 231
Colluvium 131, 217
Colombia 140, 141
Colorado 45, 100
Columbia River 144, 145, 150
Column, basalt 60, 61, 260, 264, 265
Como, Lake 224, 225
Cone, volcanic 254–257
Conglomerate 80, 81, 250
Consolidation 172, 173
Constructive plate boundary 180
Continuous deformation 28–36
Contrition 71–77
Cooling joints 264, 265
Coral 170, 171, 202–205
Corfu 206, 207
Corrasion 76, 77
Coulee 144, 145
Couloir 48
Crags on plateau 50
Crater 62, 246, 247, 254, 256–260, 270
Creep 112–126
Cretaceous 84, 86, 87, 89, 261
Cross beds 84, 85, 89
Crustal shortening 45
Crystalline rocks 12, 13, 20, 21
Currents 179, 188, 189, 208, 210
Cycle, landscape 1, 2

Dacite 262, 263
Danube 137, 154–156
Daranak falls 152, 153
Davis 1, 2, 275
Dead ice 236, 237
Debris flow 65, 270
Debris slide 128, 129
Deccan Traps 20, 21, 104, 105
Deep sea deposits 84–87
Deewhy Bay 208, 209
Deformation, continuous 28–36
Deformation, discontinuous 38–47
Degradation 1, 2, 5, 54
Delta 179, 210, 211

Denmark 227
Denudation rate 2
Denver 100
Deposition 90, 91, 176, 177, 227, 230, 231–233, 269
Desert 16, 17, 24, 25, 239–251
Design, tectonic, *see* Predesign
Devils Tower 260
Devonian 68, 250
Diabase 62, 63
Dieng plateau 58, 59, 64, 271, 273
Discontinuous deformation 38–47
Dolerite 68
Dolomite 28, 29
Dolostone 50, 148
Door, glacier 220
Drift 214, 215, 230
Drumheller Badlands 89, 99
Drumlin 230–232
Dry climate, *see* Arid climate
Dry Falls 150
Dunes 20, 21, 84, 85, 240–243
Dust 239

Earth pillar 98
Ebriach gorge 62
Echelon faults 44
Eddy 132, 137
Edmonton sandstone 89
Einsiedeln 96, 97
Elbow River 166, 167
Eluvium 131, 217
Emergence coast 182, 183
Endless Chain Ridge 46, 47
Endogenic processes 1, 2, 5, 95
Entlebuch 122, 123
Ejecta 62–64, 266–269
Erdmannlistein 229
Erg 240, 241
Erosion 1, 2, 5, 20–23, 30, 31, 98–105, 148, 149, 154–157, 190, 194, 206–209, 224–226, 256
Erratic block 229
Eruption 254, 256, 258, 266, 269, 271
Estuary 179, 193, 210–212
Ethiopia 62, 63, 266, 267
Eustatic changes 182, 183
Evaporation 239, 246
Evaporite 83, 239, 245–248
Exfoliation 52, 53
Exogenic processes 2, 57, 74
Extrusion 68, 69

Falaises 190, 191
Fast movements 127–130
Fault 42–44, 68
Ferdenpass 234, 235
Field, lava 258, 259, 266, 267
Fissure 122, 123
Flats, evaporite 248
Flood plain 140, 141, 160, 166, 167
Florida 186, 187
Flow fold 32, 33
Flow, lava 262–267
Flow, river 132, 133
Flow, shooting 5
Flow, streaming 5
Flow, subcritical 132
Flow, supercritical 133
Flow, volcanic 58–63
Fluting 86, 87, 174, 175
Fluvio-volcanics 66, 67
Flysch 86–88, 112, 113, 118, 119, 136
Fold 28–33, 144, 145
Fossil 36, 80, 81
France 70, 71, 74, 75, 190, 191
Frank 128, 129
Freezing 71
Frost polygon 230, 231
Fuorcla Surlej 110
Fyord 179, 212

Galveston 182, 183
Gansu 162, 163, 246–248
Garlands of heads 56
Garnet 92, 93
Gas, volcanic 62
Gatineau 133
Georgia 68, 69
Gerber 5, 55, 108, 111, 275
Gerber principle, *see* Selection principle
Germany 198–201
Geyser 253, 272, 273
Geyserite 273

Giessbach falls 151
Gimi 60, 61, 172, 173
Gisliflue 49
Glacial climate 11, 14, 15, 19, 22, 23
Glacial cycle 1
Glacial effects 84, 85, 90, 91, 224–237
Glacier 213, 216–221, 224–229
Glen Helen 146
Gneiss 34, 35, 40
Goose Bay 208, 209
Gorge 142–146
Gran Canaria 268
Grand Canyon 142, 143
Grand Coulee 144, 145
Granite 53, 66–69
Granulite 74
Grass slide 122, 123
Gravity effect 5, 27, 54–56
Great Erg 240, 241
Greece 206, 207, 210, 211
Greenland 213, 222, 223
Grenville 28, 29, 40, 92, 93
Ground fissure 122, 123
Ground, patterned 60, 61, 230, 231
Guadeloupe 65, 170, 171, 270
Guanare River 158, 159
Gulf coast 182, 183
Gulf stream 188, 189
Gulf, Venezuela 194–197
Gully 43, 101, 103
Gurten 104, 105
Gypsum 170, 171, 239, 245

Halite 83
Hanging glacier 221
Hard rock 12, 13, 16, 17, 20–25, 104, 105
Head, garlands 58
Head section 140, 141
Herculaneum 254
High-activity landscape 12–17
Himalaya 66, 67, 213
Hochkönig 28, 29, 82, 108, 109, 111, 172, 173, 234, 235
Hog's back 45
Honshu 265
Hoodoo 99
Hot spring 253, 274
Humid climate 1, 12–14, 16–18, 20–23
Hump 234, 235
Hyderabad 38, 39
Hydrothermal effects 272–274
Hypsometric curve 2–4

Ice 11, 90, 91, 213, 216, 218, 220–222, 224, 232, 236, 237
Ice action 11, 90, 91
Ice age 22, 90, 182, 213, 222, 224, 227, 229
Iceberg 180, 181
Ice cap 222, 223
Ice fall 218, 219
Iceland 180, 181
Ice lens 90, 91
Igneous rocks 57
Ignimbrite 62, 63
Illinois 101
Incised meander 160, 161
India 20, 21, 38, 39, 66, 67, 98, 104–106
Indonesia 14, 15, 58, 59, 64, 79–81, 138, 139, 202, 203, 254, 255, 271, 273
Indus root zone 66, 67
Inselberg 66, 69, 239, 250, 251
Instability principle 4, 5, 20, 34, 35, 104, 105, 116, 117, 136–139, 176, 214, 234
Interglacial stage 213
Intrusion 68, 69
Island 184, 185, 202, 203
Island mountain, *see* Inselberg
Italy 127, 128, 129, 190, 191, 224, 225, 236, 237, 254, 255

Jacksonville 186, 187
Japan 265
Java 58, 59, 64, 79, 80, 81, 138, 139, 254, 255, 271, 273
Joint 38–41, 74, 75, 172, 264
Jos plateau 66, 67
Jura mountains 48, 49, 107
Jurassic 52, 56, 68, 116

Kalibang formation 79
Kame 232, 233
Kar lake 224, 225
Kärselen 230, 231

Karst 172–175
Keewatin District 22, 23
Kegon falls 265
Kenang 266, 267
Kereng hill 266
Kickapoo Park 101
Klanggon 138, 139
Knauer 73
Kohlbeck 40, 62, 275
Kollbrunn 176
Kollmannsegg 117
Kufena hill 251

Labrador 208, 209
Ladakh 66, 67, 98, 106
La Florida 140, 141
Lagoon 202, 203
Lahar 65, 79, 270
Lake 28, 29, 72, 103, 117, 125, 160, 161, 179, 189, 194, 195–199, 206, 207, 218, 219, 224, 225, 232, 233, 236, 237, 246, 247, 258, 259, 270
Lake Como 224, 225
Lake Erie 148
Lake Maracaibo 72
Lake Montorfano 236, 237
Lake Ontario 148, 188, 189, 196–198, 199, 206, 207
Lake sediments 72
Lamayuru 98
Land slide 128–130
Langtang 60, 61, 261
Lanzarote 242, 243
Lanzhou 246–248
Lapilli 268, 269
Lateral moraine 228
Lateral river action 154–163
Laterite 66, 67, 72
Lava 41, 58–63, 253, 258, 262–267, 268
Leaching 250
Lecco 127, 128, 129
Lech river 76, 77, 134
Lens, ice 234
Lesser Slave Lake 194, 195
Lias 36
Limestone 42, 44, 52, 170, 171, 174, 175
Linear slopes 96, 97
Loess 162, 163, 246, 247
Longshore currents 210
Longshore drift 194
Loose materials 106–111
Low-activity landscape 20–25
Lower reach, valley 168, 169
Luzon 152, 153, 192, 204, 205, 246, 247, 256–259, 269

Machu Picchu 12, 13
Magma chamber 271
Malm 42
Malta valley 164, 165
Mangrove 200, 201
Mantle material 180
Maracaibo 72
Marine orogenesis 180, 181
Martinique 262, 263
Mass movements 112–130
Matabungkay 204, 205
Matterhorn 54
Maturity, landscape 1–3
Meander 131, 158–163
Medium-activity landscape 16–21
Merapi mountain 138
Mesa 20, 21, 100
Mesozoic 84, 85
Metamorphics 16, 17, 57, 90–93
Metasediments 90–93
Metavolcanics 92, 93
Mexico 58, 59, 256–259
Microcatena 166, 167
Middle reach, valley 166, 167
Midocean ridge 180, 253
Migmatite 32, 33, 92, 93
Milagro formation 72
Mississippi River 168, 169
Missouri 53, 168, 169
Molasse 73, 80, 81, 104, 105, 126
Monocline 30, 31
Monolith 250
Montana 164, 165
Montorfano, Lake 236, 237
Moose River 210, 211
Moraine 227, 228, 236
Mountain fracture 124–126
Mountain peak 5

Mountain stream, 76, 77
Mountain torrent 140–141
Movement, fast 127–130
Movement, slow 112–126
Mud flat 198, 199
Mud ripples 200, 201
Mungo lake 20, 21, 103
Murrumbidgee River 166, 167
Muskoka 90, 91
Mythen 12, 13

Nagelfluh 80, 81
Nasugbu 192
Natural bridge 162, 163
Near-shore circulation system 192, 194, 196, 208, 210, 212
Nevado de Toluca 258, 259
New South Wales 30, 31, 55, 103
Niagara 50, 51, 148, 149
Niche 41
Nigeria 60, 61, 66, 67, 172, 173, 251, 261, 266, 267
Nitrate 246
Niveal features 213–237
Normal slope 96, 97
Normandy 70, 71, 74, 75, 190, 191
North America 22, 23, 213, *see* also Canada, United States, Mexico
Northern Territory, Australia 20, 21, 24, 25, 146, 250
North Sea 198, 199
Northwest Territories, Canada 22, 23, 223
Nova Scotia 182, 183, 186, 187

Ocean bottom 180, 181
Ocean current 179, 188, 189
Old age landscape 1–3
Old Faithful 272
Ontario 28, 29, 32, 33, 40, 50, 51, 84, 85, 90–93, 148, 149, 188, 189, 206, 207, 210, 211, 232, 233
Opalinus clay 116
Ordovician 30, 31, 38, 39
Orogenesis 180, 181, 253
Orogenic belt 253
Ötscher 147
Oxbow lake 160, 161

Padale 20, 38, 104, 276
Paint pot 274
Painted Desert 160, 161
Paleozoic 32, 33, 90, 91, 213, 250
Parry Sound 92, 93
Patagonia 213
Patterned ground 230, 231
Pebbles 156
Penck 96, 275
Peneplanation 68
Penninic trough 86, 87
Periadriatic lineament 62, 63
Periglacial features 213, 224–237
Permian 37, 83
Peru 12, 13
Petite Terre 202, 203
Petrofabrics 36, 37
Petrology 57–93
Philippines 152, 153, 192, 204, 205, 246, 247, 256–259, 269
Physical reduction 71–75
Piedmont plain 236, 237
Pillar, earth 98
Pillow lava 62, 63
Pinch-and-swell structure 34, 35
Pingo 234, 235
Piton 262, 263
Piz Ot 43
Plain 2, 22–25, 236, 237
Plate, tectonic 180, 181
Plateau 50, 56, 172–175
Plateau edge 50, 56
Platform shore 192, 193, 204, 205
Pleistocene 213, 224
Pliocene 31, 31
Plug, volcanic 260–263
Plutonics 57, 66–69
Polygon 60, 61, 230, 231, 260, 264, 265
Pompeii 254
Pool 136
Popocatepetl 256, 257
Portuguese Bridge 266, 267
Prairie mound 232, 233
Precambrian 32, 33
Predesign, tectonic 27 ff., 43, 74, 75, 131, 168, 169

Principle of antagonism, *see* Antagonism principle
Principle of instability, *see* Instability principle
Process-response 4
Proterozoic 250
Pyroclastic sill 152, 153

Quebec 30, 31, 38, 39, 90, 91, 133, 168, 169, 193

Random patterns 4, 101, 102
Rapids 152, 153
Rax 174, 175
Recession, slopes 95–97, 101
Reduction 37, 57, 70–78
Reduction spots 37
Reef 202–205
Repose angle 106
Resonance effect 137, 200, 201
Response to stress 27
Riehl 275
Riffles-and-pools 136
Rift area 266, 267
Rill 102, 174, 175
Ripple 60, 61, 132, 137, 196, 197, 200, 201, 214, 215, 240, 241, 244
River 131–177, 240, 266, 267
River bed 136–139
River bed processes 131–139
River contrition 76, 77
River flow 132, 133
River mouth 179, 210–212
Roche moutonnée 235–237
Rocher du diamant 262
Rock 138, 139, 154, 155, 275
Rock, desert 250, 251
Rock fall 127–129
Rocky Mountains 16, 17, 45–47, 226
Root zone 66, 67
Ropy lava 58, 59
Rundhöcker 234–236
Rundle mountain 46, 47

Saguaro 249
Saguenay River 168, 169
Sahara 24, 25, 76, 77, 240–245, 249
Saint Lawrence 36, 39, 90, 91, 193
Salt 83, 239
Samsø 227
Sand 79, 82, 154, 155, 194, 196, 197, 208, 239–241
Sand bar 182, 183
Sandstone 148, 242, 243
Sangiran done 79–81
Sanur Beach 202, 203
Sattel 112
Saw-toothed ledge 49
Scarborough Bluffs 188, 189, 206, 207
Scarp, erosion 148, 149
Scheidegger 2–5, 20, 28, 38, 40, 55, 96, 99, 101, 104, 108, 111, 117, 128, 131, 157, 188, 242, 258, 262, 264, 268, 270, 275, 276
Schinznach slide 116, 120
Schist 84–87, 157
Schleswig-Holstein 198–200, 201
Schraubach 157
Scour hole 86, 87, 138, 139
Scree cone 111
Scree slope 106–110
Sea shore 70, 71
Sea surface 180, 181
Sea weed 188, 189
Secondary deposits, volcanic 65–67
Sedimentary rocks 12, 13, 18, 57, 79–91
Seebenalp 236, 237
Selection principle 5, 27, 54–56, 256
Self-gravitational effects 27, 54–56
Shale 82, 83, 148
Shear zone 44
Shield, Canadian 40, *see* also Grenville
Shingle 194, 195
Shoaling waves 186, 187
Shooting flow 5
Shore erosion 206–209
Shore platform 192, 193
Shoshone River 144, 145
Siberia 213
Sieving effect 108, 109
Sill 152, 153, 265
Silt 79
Silurian 50, 148
Sinamaica 200, 201

Sink hole 170–175
Slate 36
Slave Lake 194, 195
Slide 116, 118, 120–123, 128–130
Slip 106, 243
Slope 6, 95–130
Slow motion 112–126
Slump 112–126
Snow 213–216
Soft rock 14, 15, 22–25, 104, 105
Soil drag 118, 119
Sole marks 86, 87
Solution effects 134, 170–175
Sorting 196
Spanegg 174, 175
Spilite 62, 63
Spit 208, 209
Spontaneous mass movement 106–130
Spring, hot 253, 272, 274
Spring tufa 176
S-slope 96, 97
Stalactite 177
Stambach 140, 141
Stein glacier 14, 15, 218, 219, 224, 225
Steinlimi glacier 217–219
Stone Mountain 68, 69
Storm center 184
Strahler 2, 3, 276
Streaming flow 5
Stress, tectonic 5, 27, 37, 38
Strihen 52
Structural background 27–56, 208, 212
Structure, sedimentary 84–91
Subcritical flow 132
Submergence coast 182, 183
Sulfate 177
Sundoro volcano 254, 255
Supercritical flow 133
Surf 186, 187, 190, 202, 203
Surface effects 48–53
Suru valley 106
Susitna River 160, 161
Switzerland 12–15, 34–37, 42–44, 48, 49, 52, 54, 56, 73, 80, 81, 84–87, 96, 97, 104, 105, 107, 110, 112, 116, 120–123, 125, 126, 151, 157, 174–176, 217–219, 224, 225, 228, 230, 231, 234–237
Syncline 28
Systematic landscape features 4, 5

Taal volcano 246, 247, 256–259, 269
Tabaro 256, 257, 269
Taconian orogeny 30, 31, 38, 39, 90, 91
Tanay River 152, 153
Tasmania 68, 69, 213
Taxonomy, landscapes 1 ff., 6, 7 ff.
Taylor instability 4
Teapot effect 99
Tear scar 118–121
Tectonic background 27 ff.
Tectonic effects 28 ff.
Tectonic folds 28–31
Tectonic plate 180
Tectonic predesign 5, 74, 75, 131, 151, 168, 169, 190
Tectonic process 1, 2
Tectonic stress field 5
Teide 262, 263
Tenerife 41, 262–264
Tennessee 18
Terjung 4, 276
Terminal moraine 227
Terrace, river 162, 163
Terracettes 122, 123, 176
Texas 182, 183
Thermal process 253
Thiamis River 210, 211
Thixotropy 88
Thom 4, 276
Three Sisters 55
Thrust sheets 45–47
Tide 70, 208
Till 232
Toas, Isla 102
Toluca 258, 259
Tongue, glacier 218, 219, 227
Tor 251
Torrent 140, 141
Trachite 261
Tränenbach 170, 171
Tree 113
Trench 124
Triassic 28, 29, 82
Truncation 89

Tufa 176
Tuff 64, 79, 172, 173
Turbidity current 84–87
Turbulence 132, 240

Uetliberg 121
United States 16–18, 45, 53, 100, 101, 142, 143, 144, 145, 150, 160–165, 168, 169, 182, 183, 186, 187, 216, 249, 260, 272, 274
Uplift rate 2
Upper reach, valley 164–167
Urubamba valley 12, 13
U-valley 224, 225

Valley 164–169, 217, 224, 225
Venezuela 72, 102, 114, 115, 130, 158, 159, 194–197, 200, 201, 240, 241
Vent 260
Vertical river action 140–153
Vesuvius 254, 255
Vienna Woods 86–88, 113, 118, 119, 132, 136, 158, 159, 214, 215
Volcanic bomb 253, 266–268
Volcanic effects 152, 153, 262–271
Volcanic features 253–274
Volcanic flow 58–63
Volcanic plug 260–263
Volcanics 57, 58–67, 92, 93, 172, 173, 242, 246, 247, 270
Volcano 253, 254–259, 260

Wadden areas 198, 199
Wall effects 27, 48–51, 107
Wase Rock 261
Washington State 144, 145, 150
Water 11, 179–212
Waterfall 147–152, 218
Wave 179, 184–187, 194, 196, 214, 240, 244
Wayne 232, 233
Weathering 53, 57, 70–78, 250
Wellington mountain 68, 69
Werfen layers 82
West Indies 65, 170, 171, 184, 185, 202, 203, 262, 263, 270
White water 133
Wind 11, 20, 76, 77, 214, 239, 240
Wörschach 118, 119
Worthington glacier 216
Wyoming 144, 145, 162, 163, 260, 272, 274

Yellowstone 144, 145, 272, 274
Youth, landscape 1–3

Zeiher Homberg 48
Zuetribistock 42